Springer Texts in Education

More information about this series at http://www.springer.com/series/13812

Andy Liu

S.M.A.R.T. Circle Minicourses

 Springer

Andy Liu
Professor Emeritus
Department of Mathematical and Statistical
    Sciences
University of Alberta
Edmonton, AB
Canada

ISSN 2366-7672   ISSN 2366-7680   (electronic)
Springer Texts in Education
ISBN 978-3-319-71742-5   ISBN 978-3-319-71743-2   (eBook)
https://doi.org/10.1007/978-3-319-71743-2

Library of Congress Control Number: 2017959914

Printed on acid-free paper

This Springer imprint is published by Springer Nature
The registered company is Springer International Publishing AG
The registered company address is: Gewerbestrasse 11, 6330 Cham, Switzerland

This book is dedicated to my comrades in the outreach front line:

Maria Falk de Losada

Nikolay Konstantinov

József Pelikán

Zonghu Qiu

Wen-Hsien Sun

Peter J. Taylor

Paul Vaderlind

Preface

A Brief History of the S.M.A.R.T.Circle

A most beneficial side effect of the collapse of the former Soviet Union in 1992 was the migration of the Mathematical Circles across the Atlantic to the United States. Mathematical Circles, originated in Hungary during the nineteenth century, are a glorious tradition in Eastern Europe. They are organizations which discover and nurture young mathematical talents through meaningful extra-curricular activities.

The process took a few years, leading to the formation in 1998 of the Berkeley Mathematical Circle. With the support of the Mathematical Sciences Research Institute, the movement has caught fire in the United States, culminating in the formation of a Special Interest Group in the Mathematical Association of America under the leadership of Tatiana Shubin of San Jose State University.

Unbeknown to this community, a Mathematical Circle had existed in North America almost two decades earlier. The ultimate inspiration was still of Soviet origin, but the migration took place across the Pacific, via the People's Republic of China in the form of their Youth Palaces. This was the S.M.A.R.T. Circle in Edmonton, Canada, founded in 1981. The acronym stood for Saturday Mathematical Activities, Recreations & Tutorials.

I was born in China during the over-time sudden-death period of the Second World War, but moved to Hong Kong at age six. Thus I had never attended any session of any Youth Palace. However, I followed reports of their activities, and this fueled my interest in mathematics. The first mathematics book I had was a Chinese translation of Boris Kordemski's *Moscow Puzzles*, which was on their recommended reading list. An English version is now an inexpensive Dover paperback. Later, I acquired Chinese translations of several wonderful books by Yakov Perelman. Dover has published his *Figures for Fun* in English.

I came to Canada at age twenty, and eventually got a tenure-track position at the University of Alberta in 1980. That fall, I was invited to a general meeting of the Edmonton Chapter of the Association for Bright Children. My comment was that their activities seemed heavily biased towards the Fine Arts. Having put my foot in my mouth, I was obliged to take some concrete action. The next spring, the S.M.A.R.T. Circle was born.

During the first year, the members ranged from Grade 3 to Grade 6, because of the clientele of the A.B.C. However, to do meaningful mathematical activities, I preferred the children to be a bit more mature. So the Grade level rose by one each year, until in 1985, the members ranged from Grade 7 to Grade 10. Many of them stayed throughout this period.

As we moved away from the normal age of the clientele of the A.B.C., the Circle practically became an independent operation. This also became necessary because in 1983, we received a grant of $1,500 from the University of Alberta, arranged by Vice-President Academic **Amy Zelmer**. With the money, I built up a Circle Library containing mathematical books, games and puzzles. This was the only funding the Circle had received in its thirty-two year history.

We met on the University of Alberta campus from 2:00 pm to 3:00 pm every Saturday in October, November, February and March. A second class-room adjacent to the meeting room was open from 1:30 pm to 3:30 pm as the Circle Library. **Adrian Ashley**, a former Circle member, was hired at $5 an hour to look after it. There was a comedy of error in that for a while, his salary came out of the Student Union cafeteria account! They soon put a stop to that, but never bothered to claim readjustment.

Because of the members' tender ages, most came with their parents, and some parents stayed in Circle Library during the session. Members also had half an hour before and half an hour after the session to browse through. Sometimes, some younger members' attention span wandered during the session, and they would drift to the Circle Library for a few minutes.

In 1986, the three-year period of the grant ran out. As I closed the account, I turned the Circle Library over to the Faculty of Education. Then I started building a replacement out of my own pocket. Meanwhile, the A.B.C. had acquired new headquarters in the form of a house, where the basement was set up as a classroom. The Circle was invited to move its operation there. As a result, I restarted the session for A.B.C. members from Grade 3 to Grade 6 again. This went from 1:00 pm to 2:00 pm while the existing session for the older children ran from 2:30 pm to 3:30 pm. We had quite a few sibling pairs. Sometimes, one was in class in the basement while the other waited upstairs and played with mathematical games and puzzles from the new Circle Library. Sometimes, they sat in the same session despite any disparity in age.

In 1991, this arrangement came to an end, and the Circle moved back to the university campus. Only the Grade 7 to Grade 10 session survived the move. The meeting time was once again from 2:00 pm to 3:00 pm. A section at the back of the classroom was reserved for the Circle Library.

In 1996, there was a reverse migration of the Circle movement back across the Pacific, to Taiwan. My friend Wen-Hsien Sun of Taipei started the Chiu Chang Mathematical Circle, initially based on my model and using much of the material I had accumulated over a decade and a half. Both Circles closed in 2012, though mine was reincarnated as the J.A.M.E.S. Circle, standing for Junior Alberta Mathematics for Eager Students. It is run by my former student **Ryan Morrill**.

The activities of the S.M.A.R.T. Circle may be loosely classified into the following overlapping categories:

1. **Mathematical Conversations**;

2. **Mathematical Competitions**;

3. **Mathematical Congregations**;

4. **Mathematical Celebrations**.

The first two activities are the heart and soul of our Circle right from the beginning. The last two did not emerge until the second half of our Circle's thirty-two year history. For a description of these activities, see the companion volume *The S.M.A.R.T. Circle — Overview*.

The Mathematics Conversations are the heart and soul of the Circle. There is a Fall Session and a Winter Session each academic year. The Fall Session runs in October and November while the Winter Session runs in February and March. We meet every Saturday during those months from two to three in the afternoon at the University of Alberta. Each session consists of a minicourse plus a number of investigation topics. The latter lead to projects by Circle members, either independently or in small groups.

The material in this book is based on the best received minicourses offered over the thirty-two year history of the Circle, both in the Mathematical Conversation sessions and in the summer camps under Mathematical Congregations.

Andy Liu,
Edmonton, Alberta, 2017.

Acknowledgement

I am very excited that **Springer-Verlag**, an institution in mathematics publishing, agrees to publish this volume. I am most grateful to their staff for encouragement and support, in particular, to Jan Holland, Bernadette Ohmer and Anne Comment. The technical team of Suganya Manoharan at Scientific Publishing Services, Trichy, India, has made significant contributions to the layout of the book.

Acknowledgement

Table of Contents

Part II.   Other Topics

Part I: Geometric Topics

Chapter One: Area and Dissection

Section 1. Qualitative and Quantative Treatments of Area

We shall assume that the readers are familiar with simple terms such as triangles, squares, rectangles and parallelograms, simple concepts such as congruence, similarity, convexity and parallelism, and simple results such as the sum of the angles of a triangle being 180°. On the other hand, we will make precise the assumptions about area.

Our focus is on the area of polygons. The first assumption is called the **existence** of area, which states that *every polygon has a non-negative area*. This may seem unnecessarily legalistic, but it provides us with a context. With that in mind, we will safely ignore it from now on.

The next two assumptions allow us to give a qualitative treatment of area.

The Principle of Conservation of Area.
If a plane figure is dissected into several pieces, then its area is equal to the sum of the areas of the pieces.

The Principle of Preservation of Area.
If a plane figure is transferred to another location by rigid motion, then its area is unchanged.

In other words, congruent figures have equal area.

We start with a simple example. When a parallelogram is divided by either diagonal into two triangles, they have equal area because they are congruent.

Let E be a point on the side AB of a parallelogram $ABCD$. Is the area of triangle CDE greater than, equal to or less than one half the area of $ABCD$?

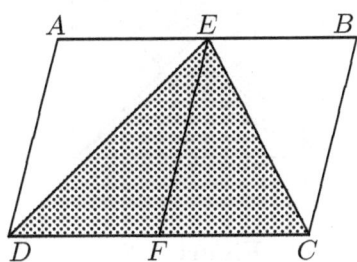

Figure 1.1

© Springer International Publishing AG 2018
A. Liu, *S.M.A.R.T. Circle Minicourses*, Springer Texts in
Education, https://doi.org/10.1007/978-3-319-71743-2_1

Draw a line through E parallel to AD, cutting CD at F, as shown in Figure 1.1. Then EF divides $ABCD$ into two parallelograms $AEFD$ and $BEFC$. Now triangle DEF has half the area of $AEFD$ while triangle CEF has half the area of $BEFC$. By the Principle of Conservation of area, triangle CDE has half the area of $ABCD$.

P is a point inside a square $ABCD$. Which if either have greater total area, triangles PAD and PCB, or triangles PAB and PCD?

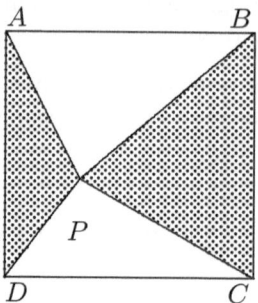

Figure 1.2

Draw a line through P parallel to AD, cutting AB at E and CD at F, as shown in Figure 1.2. Then PAD has half the area of $ADFE$ and PCB has half the area of $BCFE$. Together they have half the area of $ABCD$, and their total area is equal to the total area of PAB and PCD.

P and R are points inside a rectangle $ABCD$ such that RA cuts PB at Q and RD cuts PC at S, as shown in Figure 1.3. Prove that the total area of triangles PAQ and PDS is equal to the total area of triangles RBQ and RCS.

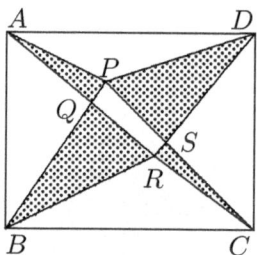

Figure 1.3

When triangles ABQ and CDS are added to both side, the augmented total area of each side is now half the area of $ABCD$. Hence the total area of each side before augmentation must also be the same.

Up to now, we have presented a qualitative treatment of area. Now we move on to a quantitative treatment. To that end, we choose the rectangle as our basic figure and accept as the basic formula how its area is to be computed. We then compute the areas of other polygons based on this formula.

The Basic Formula of Area.

The area of a rectangle as the product of its length and its width.

From this basic formula, we can derive formulas for the areas of various polygons.

Theorem 1. The area of a square is equal to the square of its side.

Proof:

This is because a square is also a rectangle. Its side is both the length and the width of the rectangle. The desired result follows from the basic formula.

Note in particular that the area of a 1×1 square is indeed 1.

Theorem 2. The area of a parallelogram is equal to the product of its base and its height.

Proof:

Cut the parallelogram into two pieces along the perpendicular from a vertex to the longer side. Then the foot of the perpendicular is on the side. Re-assemble the two pieces into a rectangle, as shown in Figure 1.4. Because of the conservation and preservation of area, the parallelogram has the same area as the resulting rectangle. Now the parallelogram's base is equal to the rectangle's length, and the parallelogram's height is equal to the rectangle's width. The desired result follows from the basic formula.

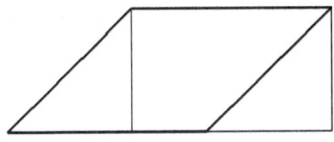

Figure 1.4

Theorem 3. The area of a triangle is equal to half the product of its base and its height.

First Proof:

Put two copies of the triangle together to form a parallelogram, as shown in Figure 1.5 on the left. By the Principles of Conservation and Preservation of area, the area of the triangle is half that of the resulting parallelogram. Now the triangle's base and height are equal to that of the parallelogram. It follows from Theorem 2 that the area of the triangle is equal to half the product of its base and height.

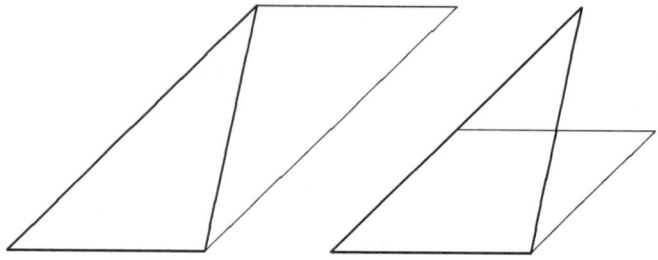

Figure 1.5

Second Proof:
Cut the triangle into two pieces along a line joining the midpoints of two sides, and reassemble them into a parallelogram, as shown in Figure 1.5 on the right. By the Principles of Conservation and Preservation of area, the triangle has the same area as the resulting parallelogram. Now the triangle's base is equal to that of the parallelogram, and the triangle's height is double that of the parallelogram. The desired result follows from Theorem 2.

Finally, the area of an arbitrary polygon can be obtained by dissecting it up into triangles, computing the area of each triangle, and applying the Principle of Conservation of Area.

If the polygon is convex, the dissection is simple. We just connect any vertex to all the other vertices, as shown in Figure 1.6 on the left. Suppose the polygon is non-convex. We can eliminate an angle greater than a straight angle extending one of its sides beyond the vertex, until it hits the perimeter of the polygon again, as shown in Figure 1.6 on the right. Eliminating such angles one at a time, we have dissected the polygon into convex polygons. We can now deal with each of them as before.

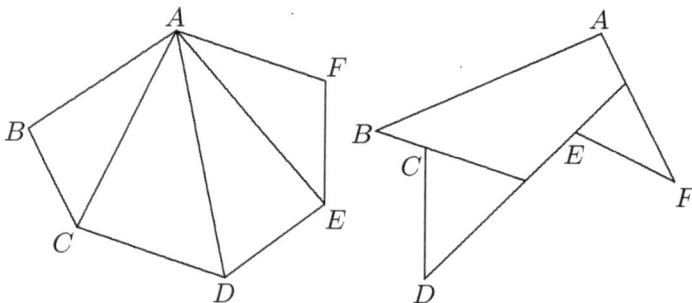

Figure 1.6

It would appear that the 1×1 square is the more natural choice as the basic figure. We then assign it an area of 1. All we need to do is to derive that the area of an $\ell \times w$ rectangle is ℓw. Then we can pick up where we left off.

Suppose both ℓ and w are integral, say $\ell = 4$ and $w = 3$. Cut the 3×4 rectangle into $3 \times 4 = 12$ unit squares. Since each square has area 1, the rectangle has area 12 by the Principle of Conservation of Area.

Suppose only one of ℓ and w is integral, say $\ell = 4$ and $w = \frac{3}{5}$. Use 5 copies of it to form a 3×4 rectangle. We have seen that this enlarged rectangle has area 12. Since it is composed of 5 copies of the original rectangle, the area of the original rectangle is $\frac{12}{5}$, again by the Principle of Conservation of Area.

Finally, suppose neither ℓ nor w is integral, say $\ell = \frac{4}{7}$ and $w = \frac{3}{5}$. Use 7 copies of it to form a $\frac{3}{5} \times 4$ rectangle. We have seen that this enlarged rectangle has area $\frac{12}{5}$. Since it is composed of 7 copies of the original rectangle, the area of the original rectangle is $\frac{12}{35}$.

It would appear that it poses no problem if we take the unit square as the basic figure for area measurement rather than the rectangle. The situation is actually quite a bit more complicated. For instance, What is the area of a $1 \times \sqrt{2}$ rectangle? It should be $\sqrt{2}$, but how are we going to justify it?

Imitating the approach above, we would like to use n copies of it to form an $1 \times m$ rectangle for some positive integers m and n. Then the enlarged rectangle has area m, so that the area of the original rectangle is $\frac{m}{n}$. However, this presupposes that $\sqrt{2}$ is expressible in the form $\frac{m}{n}$.

We shall use an indirect argument to show that this is impossible. Suppose to the contrary that $\sqrt{2} = \frac{m}{n}$ for some positive integers m and n. By canceling any common divisors of m and n, we may assume that m and n are not both even. From $2 = \frac{m^2}{n^2}$, we have $2n^2 = m^2$. Since $2n^2$ is even, so is m^2, and it follows that m must be even. Hence $m = 2k$ for some positive integer k. Now $2n^2 = m^2 = 4k^2$ so that $n^2 = 2k^2$. The same argument yields that n is also even, but this contradicts our hypothesis on m and n.

A number expressible in the form $\frac{m}{n}$ for integers m and $n \neq 0$ is called a *rational* number. At one point, the ancient Greeks believed that all numbers were rational. Along came Pythagoras who showed that the length of the diagonal of a unit square is $\sqrt{2}$, and this number turns out to be *irrational*. Their existence was not easy for the mathematicians of the day to accept, and this precipitated a major crisis in the mathematical world of ancient Greece. Eventually, the crisis was resolved by one of them, Eudoxus, who postulated the following property of irrational numbers.

The Principle of Eudoxus.

Let x be an irrational number and $\frac{m}{n}$ be any rational number. If the number y is such that $y > \frac{m}{n}$ whenever $x > \frac{m}{n}$, and $y < \frac{m}{n}$ whenever $x < \frac{m}{n}$, then $y = x$.

Let us see how this Principle helps us find the area of a $1 \times \sqrt{2}$ rectangle. Denote its area by A. Let $\frac{m}{n}$ be any rational number. Then the area of the $1 \times \frac{m}{n}$ rectangle is $\frac{m}{n}$. If $\sqrt{2} > \frac{m}{n}$, then the $1 \times \sqrt{2}$ rectangle is larger than the $1 \times \frac{m}{n}$ rectangle, so that $A > \frac{m}{n}$. Similarly, if $\sqrt{2} < \frac{m}{n}$, then $A < \frac{m}{n}$. By the Eudoxian Principle, $A = \sqrt{2}$. In other words, the area of the $1 \times \sqrt{2}$ rectangle is $\sqrt{2}$.

In the preceding paragraph, we may replace $\sqrt{2}$ by any irrational number w. Hence the area of a $1 \times w$ rectangle is w. It is now easy to show that the area of a $\ell \times w$ rectangle is ℓw if ℓ is rational. As an illustrative example, consider a $3 \times w$ rectangle. Cut it into 3 $1 \times w$ rectangles, each of which has area w. Hence the original rectangle has area $3w$. As another illustrative example, consider a $\frac{3}{5} \times w$ rectangle. Use 5 copies of it to form a $3 \times w$ rectangle, which has area $3w$. Hence the original rectangle has area $\frac{3w}{5}$.

What about if neither ℓ nor w is rational? Denote by A the area of an $\ell \times w$ rectangle. Let $\frac{m}{n}$ be any rational number. Then the $\frac{m}{n} \times w$ rectangle has area $\frac{mw}{n}$. If $\ell > \frac{m}{n}$, then the $\ell \times w$ rectangle is larger than the $\frac{m}{n} \times w$ rectangle, so that $A > \frac{mw}{n}$ or $\frac{A}{w} > \frac{m}{n}$. Similarly, if $\ell < \frac{m}{n}$, then $\frac{A}{w} < \frac{m}{n}$. By the Eudoxian Principle, $\frac{A}{w} = \ell$ or $A = \ell w$. In other words, the area of the $\ell \times w$ rectangle is ℓw.

Exercises:

1. (a) Figure 1.7 on the left shows a regular five-pointed star. How does the area of the shaded region compare with the total area of the unshaded regions?

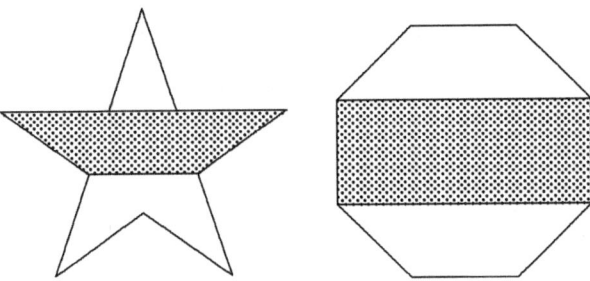

Figure 1.7

(b) Figure 1.7 on the right shows a regular octagon. How does the area of the shaded region compare with the total area of the unshaded regions?

2. (a) E is a point on the side BC and F is a point on the side CD of a rectangle $ABCD$. BF cuts AE at G and DE at K while AF cuts DE at H, as shown in Figure 1.8. Prove that the area of the quadrilateral $AGKH$ is equal to the total area of the quadrilateral $CEKF$ and the triangles BEG and DFH.

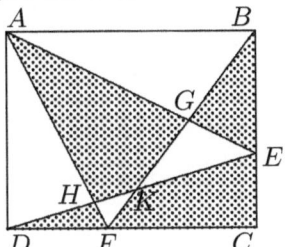

Figure 1.8

(b) From a point P inside an equilateral triangle ABC, perpendiculars are dropped to the sides BC, CA and AB at D, E and F respectively, as shown in Figure 1.9. Which if either have greater total area, triangles PAF, PBD and PCE, or triangles PAE, PCD and PBF?

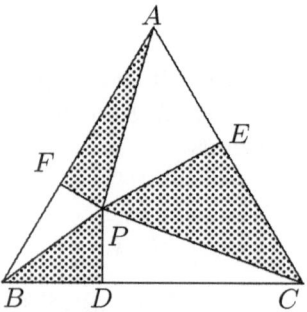

Figure 1.9

3. From a point P inside an equilateral triangle ABC, perpendiculars are dropped to the sides BC, CA and AB at D, E and F respectively. Where should the point P be

(a) if $PD + PE + PF$ is to be as small as possible;

(b) if $PA + PB + PC$ is to be as small as possible?

Section 2. The Bolyai-Gerwin Theorem and Pythagoras' Theorem.

By the Principles of Conservation and Preservation of Area, if one polygon is obtained from another by dissection and reassembly, then the two polygons must have the same area. It is natural to ask whether the converse statement is also true. In other words, if two polygons have the same area, can one always be obtainable from the other by dissection and reassembly?

With an endless variety of polygons, the task is seemingly unmanageable. We can simplify the situation somewhat by asking if any polygon can be dissected and reassembled into a square of equal area. If this is so, we can transform one polygon into another via the square.

Let us illustrate with a simple example. The first polygon is a 1×4 rectangle. The second polygon is a parallelogram with base 1 and height 1, and the acute angle is $45°$. Each can be cut into two pieces and reassembled into a square, as shown in Figure 1.10.

Figure 1.10

Superimpose one square over the other, as shown in Figure 1.11 on the left. Transferring the cuts back to the original polygons, we obtain a four-piece solution as shown in Figure 1.11 on the right.

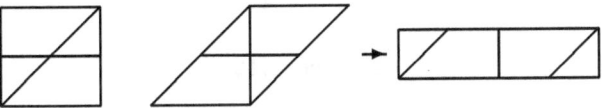

Figure 1.11

Thus we can now focus only on transforming polygons into squares.

The methods we used in developing formulas for the area of various polygons now help us establish the following remarkable result.

Theorem 5 (The Bolyai-Gerwin Theorem).
Every polygon can be dissected into a finite number of pieces which can be reassembled into a square of the same area.

Proof:

We give a constructive proof in five steps.

Step 1. Cut the polygon into triangles.

Step 2. Transform each triangle into a parallelogram.

Step 3. Transform each parallelogram into a rectangle.

Step 4. Transform each rectangle into a square.

Step 5. Combine the squares two at a time.

Step 5 is then repeated until the final square is obtained.

The first three steps have already been done, as shown respectively in Figures 6, 5 and 4 of Section 1. The last two steps require much more work.

Step 4. Transform each rectangle into a square.

We first construct a line segment of length $\sqrt{ab}$ given segments of lengths $a > b$. Construct AC with a point B on it, such that $AB = a$ and $BC = b$. Draw a semicircle with diameter AC, and erect a perpendicular from B to AC, cutting the semicircle at D, as shown in Figure 1.12 on the left. We claim that BD has the desired length.

Let O be the center of the semicircle. Then OAD and OCD are isosceles triangles, as shown in Figure 1.12 on the right. Hence $\angle OAD = \angle ODA$ and $\angle ODC = \angle OCD$. Now the sum of these four angles is $180°$. Hence $\angle ADC = \angle ODA + \angle ODC = 90°$. Then $\angle BDC = 90° - \angle BDA = \angle BAD$. It follows that triangles BAD and BDC are similar, so that $\frac{BD}{BA} = \frac{BC}{BD}$. Hence $BD^2 = BA \cdot BC = ab$ so that we indeed have $BD = \sqrt{ab}$.

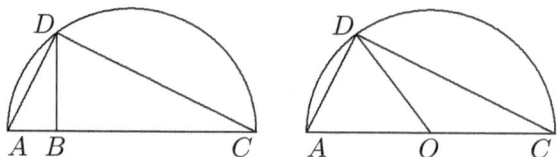

Figure 1.12

Let $ABCD$ be a rectangle with $AB = a$ and $BC = b$, where $a > b$. Let E be the point on the extension of DA such that $DE = \sqrt{ab}$. Complete the square $DEFG$ with G on CD. Let AB cuts FG at R. Join CE, cutting AB at P and FG at Q, as shown in Figure 1.13 on the left.

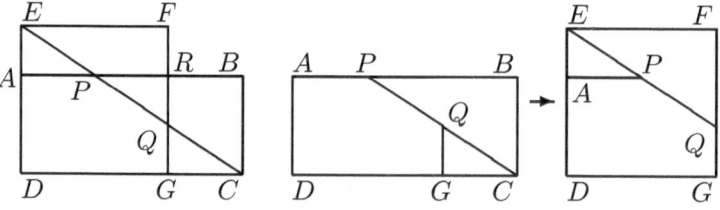

Figure 1.13

Triangles CDE and CGQ are similar to each other. Hence $\frac{GQ}{GC} = \frac{DE}{DC}$. We have $GC = b - \sqrt{ab}$, $DC = b$ and $DE = \sqrt{ab}$. Let $GQ = x$. Then $\frac{x}{b-\sqrt{ab}} = \frac{\sqrt{ab}}{b}$. Cross multiplying yields $bx = b\sqrt{ab} - ab$. Dividing by b, we have $x = \sqrt{ab} - a$. This is the length of AE. It follows that triangles CGQ and PAE are congruent, as are triangles PCB and EQF. Thus we have a dissection of $ABCD$ into three pieces which may be reassembled into $DEFG$, as shown in Figure 1.13 on the right.

Note that if $a : b = 4 : 1$, R lies on CE, and if $a : b > 4 : 1$, R will be on the same side of the line CE as D. The remedy is to cut the rectangle along the perpendicular bisector of the longer side and reassemble the two pieces into a rectangle whose dimensions are in a smaller ratio, and this is repeated until the ratio is less than 4:1.

Step 5. Combine the squares two at a time.

If the two squares are of the same size, cut each in half along a diagonal. The four isosceles right triangles so obtained can easily be assembled into a square. Henceforth, we assume that the two squares are not of equal sizes. Let the bigger one be $ABCD$, and let the smaller one $DEFG$ be placed alongside it so that G lies on AD and E lies on the extension of CD, as shown in Figure 1.14. Let H be the point on CE and K be the points on the extension of GA such that $CH = EF = AK$. Note that triangles ABK, CBH, EHF and GKF are congruent to one another, so that $BH = HF = FK = KB$. In fact, $BHFK$ is a square because we also have $\angle HBK = \angle HBA + \angle ABK = \angle HBA + \angle CBH = \angle CBA = 90°$. So we cut along BH and HF, move triangle EHF to ABK and triangle CBH to GKF.

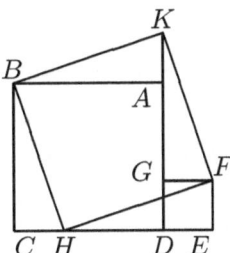

Figure 1.14

This completes the proof of the Bolyai-Gerwin Theorem.

Three country gentlemen go for a trot on a donkey, a horse and a hippopotamus, respectively. The first has a boy who weighs 30 kilograms. The second has a boy who weighs 40 kilograms. The third one weighs 70 kilograms himself.

What conclusion can one draw from this situation?

In Figure 1.14, $BHFK$ is the square on the hypotenuse FH of the right triangle EFH, $DEFG$ is the square on the side EF of EFH, while $ABCD$ may be regarded as the square on the side EH of EFH because we have $EH = CE - CH = CE - DE = CD$. Thus we have proved what is certainly one of the best known results in Euclidean Geometry.

Theorem 6 (Pythagoras' Theorem).
The square on the hypotenuse is equal to the sum of the squares on the other two sides.

In other words: *The squire on the hippopotamus is equal to the sons of the squires on the other two rides.*

There are many other ways of proving Pythagoras' Theorem. Note that its converse is also true, and may be expressed in the following way.

Theorem 7 (Pythagoras' Inequalities).
Let ABC be any triangle.

(a) If $\angle BCA < 90°$, then $AB^2 < AC^2 + BC^2$.

(b) If $\angle BCA > 90°$, then $AB^2 > AC^2 + BC^2$.

Proof:

(a) If $AB < AC$, there is nothing to prove. Hence we may assume that $AB > AC$, so that the foot D of perpendicular from A to BC lies on BC, as shown in Figure 1.15 on the left. It follows from Pythagoras' Theorem that $AB^2 = AD^2 + BD^2 < AC^2 + BC^2$.

(b) The foot D of perpendicular from A to BC lies on the extension of BC, as shown in Figure 1.15 on the right. It follows from Pythagoras' Theorem that

$$AB^2 = AD^2 + (BC + CD)^2 = AC^2 + BC^2 + 2CD \cdot BC > AC^2 + BC^2.$$

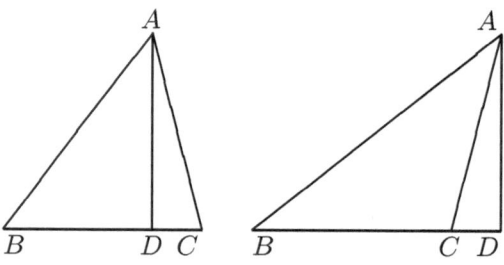

Figure 1.15

Exercises

4. Figure 1.16 shows two figures of equal area. The one on the left has three right angles and two pairs of equal sides. The one on the right has two right angles, three equal sides and two other equal sides meeting at an angle equal to three right angles. What is the minimum number of pieces into which the first has to be dissected in order to reassemble them to form the second?

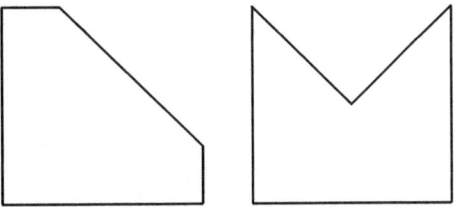

Figure 1.16

5. Figure 1.17 shows three squares $ABDE$, $BCFG$ and $CAHI$ drawn outside triangle ABC which has a right angle at C.

   (a) Prove that triangles BAG and BDC are congruent.
   (b) Which triangle is congruent to BAH?
   (c) Deduce Pythagoras' Theorem from (a) and (b).

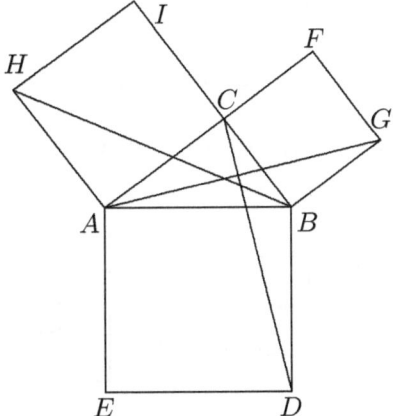

Figure 1.17

6. Derive from Pythagoras' Theorem a result known as Heron's Formula, which states that the area of a triangle is $\sqrt{s(s-a)(s-b)(s-c)}$, where a, b and c are the side lengths and $s = \frac{a+b+c}{2}$ is the semi-perimeter.

Section 3. Dissection Problems

For the dissection of one polygon to be reassembled into another of equal area, the Bolyai-Gerwin Theorem guarantees a method but not necessarily in the smallest number of pieces. What is significant is that the task can be accomplished in a finite number of pieces. Otherwise, one can melt a polygon down and pour the substance into a plate in the shape of the target polygon.

Naturally, we desire solutions of dissection problems in the smallest number of pieces. For instance, the dissection and reassembly in Figure 1.11 could have been accomplished in three pieces, as shown in Figure 1.18.

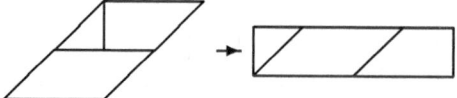

Figure 1.18

When the area of a polygon is the square of an integer, there is often an easier way to transform it into a square in a small number of pieces. For example, the polygon in Figure 1.19 has area 9.

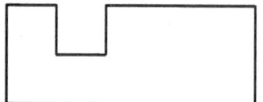

Figure 1.19

We can cut it into two pieces which can form a 3×3 square, as shown in Figure 1.20.

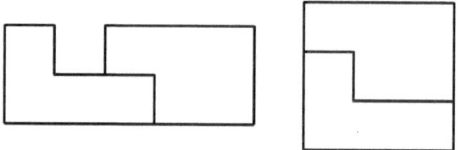

Figure 1.20

When the area n of a polygon is not an integer, the side length of the target square is $\sqrt{n}$. In the fourth step of the proof of the Bolyai-Gerwin Theorem, we have developed a general method of constructing a segment of this length. However, for specific values of n, there are easier ways to obtain segments of length $\sqrt{n}$.

For instance, a segment of length $\sqrt{2}$ may be obtained as the hypotenuse of a right triangles with legs of lengths 1 and 1. We call $(1, 1, \sqrt{2})$ a Pythagorean triple. Other Pythagorean triples of relevance are $(1, 2, \sqrt{5})$, $(2, 2, \sqrt{8})$, $(1, 3, \sqrt{10})$, $(2, 3, \sqrt{13})$, $(1, 4, \sqrt{17})$, $(3, 3, \sqrt{18})$, $(2, 4, \sqrt{20})$ and of course (3,4,5).

These triples are very useful for solving dissection and reassembly puzzles. For example, the polygon in Figure 1.21 has area 10.

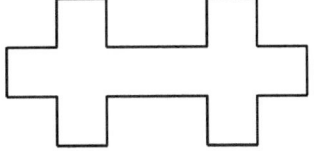

Figure 1.21

The side length of the target square is the length of a diagonal of a 1×3 rectangle. Making use of the Pythagorean triple $(1, 3, \sqrt{10})$, we obtain a four-piece solution shown in Figure 1.22.

Figure 1.22

We now tackle the problem of transforming equilateral triangles, regular hexagons, regular octagons and regular dodecagons into squares of the same area. It turns out that tessellations are most useful in solving this problem.

In Figure 1.23, the dotted lines constitute a tessellation with squares and regular octagons. The solid lines constitute another tessellation in which both tiles are squares. The smaller square is congruent to the square in the first tessellation. The larger one is therefore equal in area to the octagon. The superimposition of the two tessellations yields a five-piece dissection which transforms a regular octagon into a square of the same area. The solid lines pass through midpoints of the sides of the regular octagons.

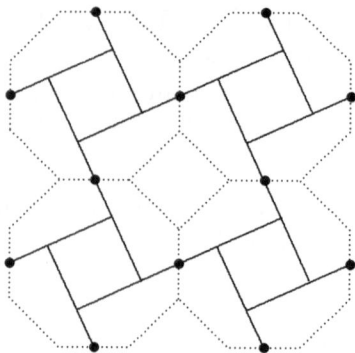

Figure 1.23

Instead of superimposing entire tessellations, we may focus on the superimposition of two tiled infinite straight strips, with their relative widths
chosen so that all tiles have the same area.

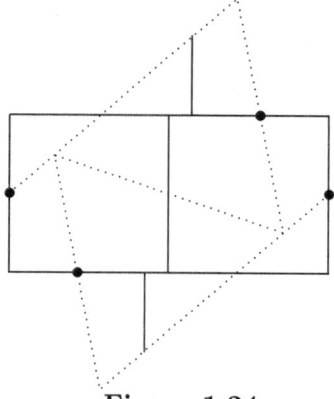

Figure 1.24

Figure 1.24 shows such a combination yielding a four-piece dissection
which transforms an equilateral triangle into a square of the same area. The
edges of the dotted strip pass through midpoints of sides of the squares while
the edges of the solid strip pass through midpoints of sides of the equilateral
triangles. This is a famous example credited to Hugo Steinhaus.

While the regular hexagon tiles the plane, it does not tile the straight
strip. However, we can cut it in halves and put the two pieces together
to form a parallelogram. In Figure 1.25, we superimpose a straight strip
tiled with such parallelograms on a straight strip tiled with squares. The
edges of the dotted strip pass through vertices of the parallelograms while
the edges of the solid strip pass through vertices of the squares. This yields
a five-piece dissection which transforms a regular hexagon into a square of
the same area.

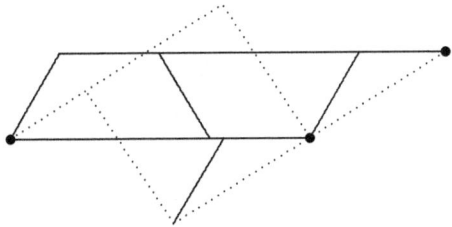

Figure 1.25

For the transformation of a regular dodecagon into a square of the same area, we cut the former into four pieces and put them together to form a parallelogram, as shown in Figure 1.26.

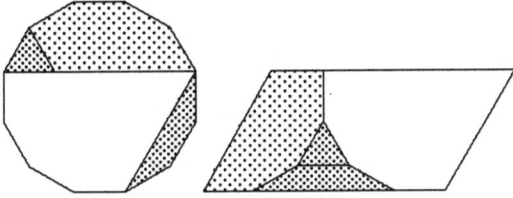

Figure 1.26

Using a straight strip tiled with this parallelogram, we obtain in Figure 1.27 a six-piece dissection which transforms a regular dodecagon into a square of the same area. The edges of the dotted strip pass through vertices of the parallelograms while the edges of the solid strip pass through vertices of the squares.

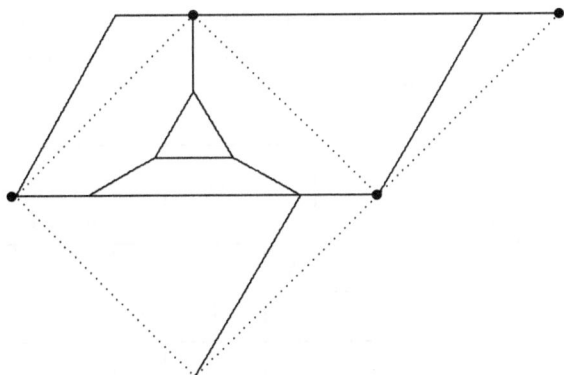

Figure 1.27

Exercises:

7. (a) Figure 1.28 shows a 7×10 rectangle with a central 1×6 rectangle missing. Dissect it into two pieces and reassemble them to form an 8×8 square.

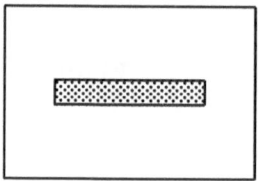

Figure 1.28

   (b) Dissect the polygon in Figure 1.29 into three pieces and reassemble them to form a square.

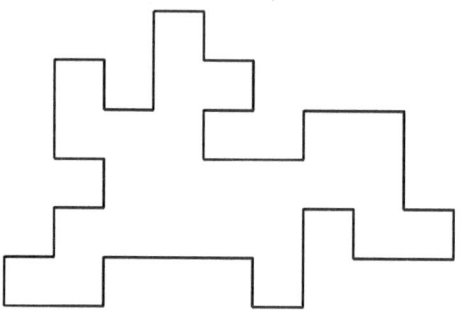

Figure 1.29

8. Figure 1.30 shows the twelve pentominoes, that is, figures consisting of five unit squares joined edge to edge. Each has a letter name. Dissect each into three or four pieces and reassemble them to form a $\sqrt{5} \times \sqrt{5}$ square.

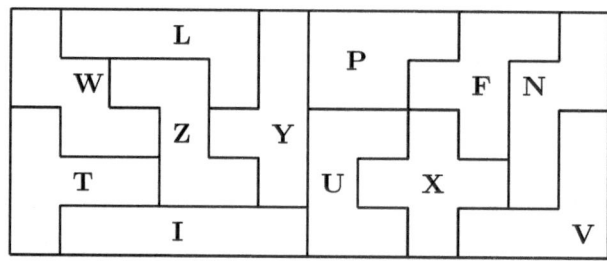

Figure 1.30

9. Figure 1.31 shows a regular pentagon cut into three pieces and re-assembled to form a parallelogram. Make use of this to find a six-piece dissection which transform a regular pentagon into a square.

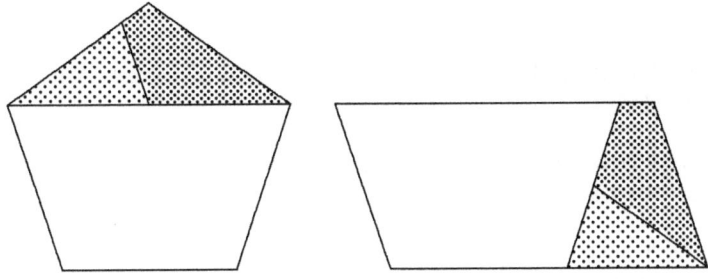

Figure 1.31

Bibliography

[1] V. G. Boltyanski, *Equivalent and Equidecomposable Figures*, D. C. Heath and Company (1963).

[2] Frederickson, Greg, *Dissections: Plane and Fancy*, Cambridge University Press (1997).

[3] Frederickson, Greg, *Hinged Dissections: Swinging & Twisting*, Cambridge University Press, (2002).

[4] Frederickson, Greg, *Piano-Hinged Dissections: Time to Fold!*, A. K. Peters Limited (2006).

[5] Gardner, Martin, Geometric dissections, in *Knots and Borromean Rings, Rep-tiles and Eight Queens*, Cambridge University Press & Mathematical Association of America (2014) 38–49.

[6] Heath, Sir Thomas L., *Euclid: the Thirteen Books of The Elements*, Vol. 1 & 2, Dover Publications Incorporated (1956).

[7] Harry Lindgren, *Geometric Dissections*, Dover Publications Incorporated (1972).

Solution to Exercises

1. (a) In Figure 1.32, each of the two parts of the star is divided into three triangles. Those on one side are congruent to the corresponding ones on the other side. Hence the area of the shaded region is equal to the total area of the unshaded regions.

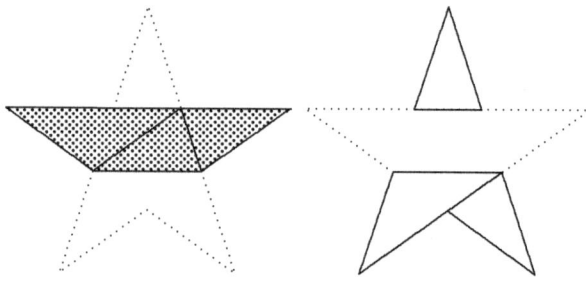

Figure 1.32

(b) In Figure 1.33, each of the two parts of the octagon is divided into two rectangles and four right isosceles triangles. The six pieces on one side are congruent to the corresponding pieces on the other side. Hence the area of the shaded region is equal to the total area of the unshaded regions.

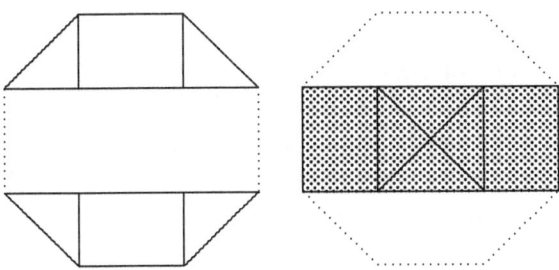

Figure 1.33

2. (a) If we add triangles BAG and FHK to the quadrilateral $AGKH$, we obtain triangle FAB, as shown in Figure 1.34 on the left. It has half the area of $ABCD$. If we add the same two triangles to the other three triangles, we obtain triangles ABE and CDE, as shown in Figure 1.34 on the right. Together, they also have half the area of $ABCD$. The desired result follows.

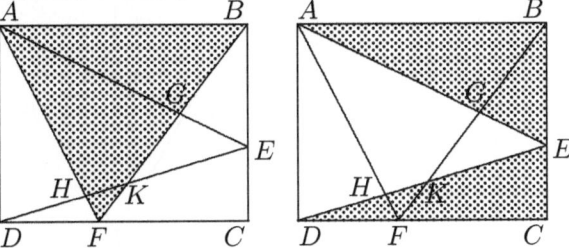

Figure 1.34

(b) Draw three lines through P parallel to the sides of ABC, as shown in Figure 1.35. They divide ABC into three equilateral triangles and three parallelograms. Exactly half of each of these six polygons is shaded. Hence the total shaded area of ABC is equal to the total unshaded area.

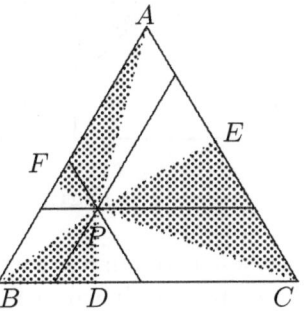

Figure 1.35

3. (a) The area of ABC is the sum of the areas of PBC, PCA and
PAB, which is respectively the product of PD, PE and PF
with the side length of ABC. Since both the area and the side
length of ABC is constant, $PD + PE + PF$ is also constant.
Hence P may be any point inside ABC.

(b) Let O be the center of ABC and P be an arbitrary point inside
ABC. Figure 1.36 shows the equilateral triangle XYZ where
A, B and C are the midpoints of YZ, ZX and XY respectively.
By (a), the sum of the perpendicular distances from O to the three
sides of XYZ is equal to the sum of those from P. The former is
just $OA + OB + OC$. Now the perpendicular distance from P to
YZ is less than PA, the perpendicular distance from P to ZX is
less than PB, and the perpendicular distance from P to XY is
less than PC. It follows that $OA + OB + OC < PA + PB + PC$,
so that O is the unique point which minimize the sum of the
distances to the vertices of ABC.

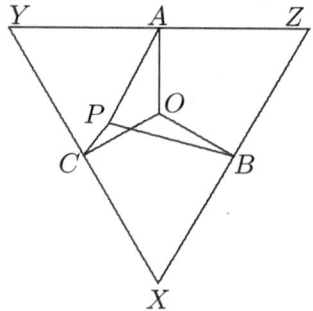

Figure 1.36

4. By adding the same isosceles right triangle to both figures, we obtain
two congruent squares, as shown in Figure 1.37.

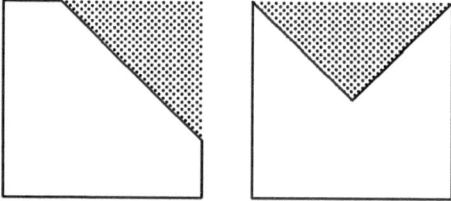

Figure 1.37

Superimposing these two squares, with one of them rotated 180°, yields
a two-piece dissection, as shown in Figure 1.38.

Figure 1.38

5. (a) We have $BA = BD$ since they are sides of the same square. Similarly, $BG = BC$. Now $\angle BAG = 90° + \angle ABC = \angle BDC$. It follows that BAG and BDC are congruent.

   (b) Similar reasoning shows that EAC is congruent to BAH.

   (c) From C, drop a perpendicular to DE, cutting AB at J and DE at K, as shown in Figure 1.39. Now BAG has half the area of the square $BCFG$ while BDC has half the area of the rectangle $BDKJ$. Hence the square $BCFG$ is equal in area to the rectangle $BDKJ$. Similarly, the square $CAHI$ is equal in area to the rectangle $AEKJ$. It follows that the total area of the squares $BCFG$ and $CAHI$ is equal to the total area of the rectangles $BDKJ$ and $AEKJ$, which is the area of the square $ABDE$.

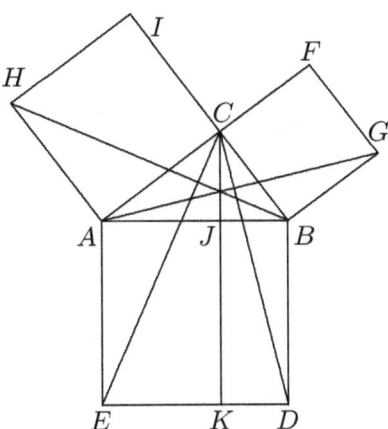

Figure 1.39

6. Let the triangle be ABC with $BC = a$, $CA = b$ and $AB = c$. Let D be the foot of perpendicular from A to BC. By taking BC to be the longest side if necessary, we may assume that D lies on BC. Let $DC = x$ and $DA = y$. Then $BD = BC - CD = a - x$. Applying Pythagoras' Theorem to triangles CAD and BAD, we have

$$b^2 - x^2 = y^2 = c^2 - (a - x)^2.$$

This simplifies to $a^2 + b^2 - c^2 = 2ax$ so that $x = \frac{a^2 + b^2 - c^2}{2a}$. Substituting this back into $y^2 = b^2 - x^2$, we have

$$
\begin{aligned}
y^2 &= (b + x)(b - x) \\
&= \left(b + \frac{a^2 + b^2 - c^2}{2a}\right)\left(b - \frac{a^2 + b^2 - c^2}{2a}\right) \\
&= \frac{a^2 + 2ab + b^2 - c^2}{2a} \cdot \frac{c^2 - a^2 + 2ab - b^2}{2a}.
\end{aligned}
$$

It follows that

$$
\begin{aligned}
4a^2 y^2 &= ((a + b)^2 - c^2)(c^2 - (a - b)^2)) \\
&= (a + b + c)(a + b - c)(b + c - a)(c + a - b).
\end{aligned}
$$

We have $a + b - c = (a + b + c) - 2c = 2s - 2c$. Hence $\frac{a+b-c}{2} = s - c$. Similarly, $\frac{b+c-a}{2} = \frac{(a+b+c)-2a}{2} = s - a$ and $\frac{c+a-b}{2} = \frac{(a+b+c)-2b}{2} = s - b$. It follows that the area of ABC is given by

$$
\begin{aligned}
\frac{ay}{2} &= \sqrt{\frac{4a^2 y^2}{16}} \\
&= \sqrt{\frac{a + b + c}{2} \cdot \frac{a + b - c}{2} \cdot \frac{b + c - a}{2} \cdot \frac{c + a - b}{2}} \\
&= \sqrt{s(s - a)(s - b)(s - c)}.
\end{aligned}
$$

7. (a) Figure 1.40 shows a two-piece solution.

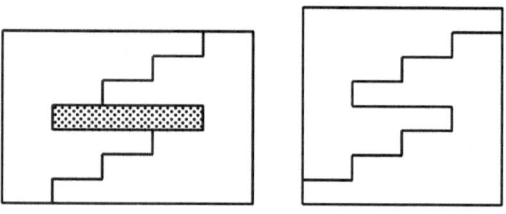

Figure 1.40

(b) Although the area of the polygon is 25, there is no three-piece dissection if we cut only along grid lines. This is a rare example in which we make use of the Pythagorean triple (3,4,5) in a dissection problem. Then we have a three-piece solution, as shown in Figure 1.41.

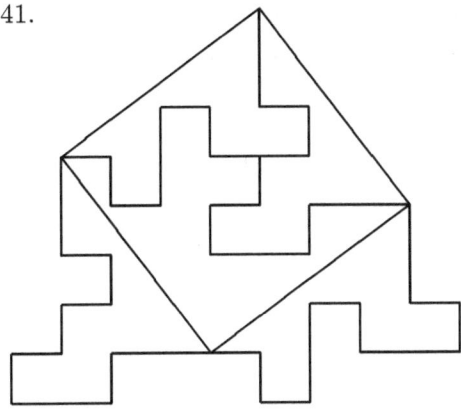

Figure 1.41

8. The area of each polygon is 5. Hence The side length of the target square is the length of a diagonal of a 1×2 rectangle. Making use of the Pythagorean triple $(1, 2, \sqrt{5})$, we obtain a three-piece solution for each of the P-, Y- and Z-pentominoes in Figure 1.42.

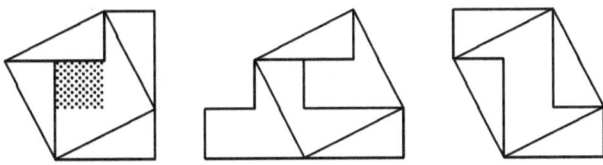

Figure 1.42

If we cut the shaded unit square from the P-pentomino and reposition it, we can form the F-, L-, N-, T-, U- and V-pentomino. Thus we have a four-piece solution for each of these six pentominoes. We obtain a four-piece solution for each of the I-, W- and X-pentominoes in Figure 1.43.

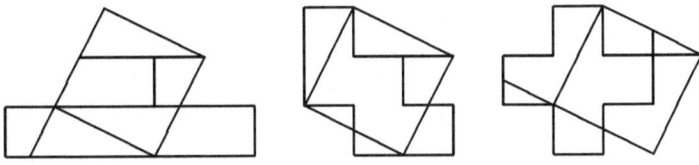

Figure 1.43

9. In Figure 1.44, we superimpose a straight strip, tiled with parallelograms converted from regular pentagons, on a straight strip tiled with
squares. This yields a six-piece dissection which can transform a regular pentagon into a square of the same area. The edges of the solid
strip pass through vertices of the squares.

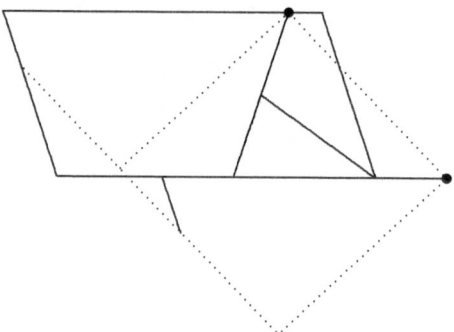

Figure 1.44

Chapter Two: Projective Geometry

Section 1. Synthetic Approach.

We begin with projections between two coplanar lines.

Let ℓ and ℓ' be two lines intersecting at a point F, and V be a point not on either of them. A projection from V of ℓ to ℓ' may be defined as follows. For each point P on ℓ, draw the line VP. If this line intersects ℓ' at a point P', we say that P' is the image of P under the projection from V. In particular, $F' = F$.

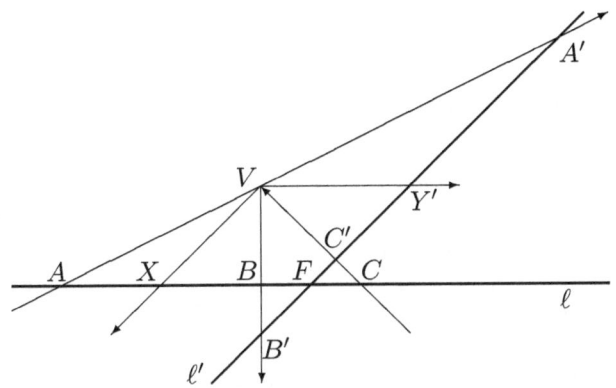

Figure 2.1

Note that the point X on ℓ where VX is parallel to ℓ' does not have an image. Also, the point Y' on ℓ' where VY' is parallel to ℓ is not the image of any point on ℓ. A simple and simple-minded remedy is to define Y' as the image of X, but this is not a satisfactory solution. We want projections to be **continuous** transformations. In other words, if a point P moves along ℓ towards another point Q, we want P' to move along ℓ' towards Q'.

Let B and C' be the respective feet of perpendicular from V to ℓ and ℓ'. As P moves from C towards B, P' does move from C' towards B'. However, as P moves on towards X, P' certainly does not move towards Y' but towards an imaginary point at infinity. Such a point does not exist in Euclidean Geometry. In Projective Geometry, we make the line ℓ' into a **projective line** $\overline{\ell'}$ by adding this point at infinity. We call it the **ideal point** of $\overline{\ell'}$, and it is taken to be the image X' of X. The other points of $\overline{\ell'}$ are called **ordinary points**.

Do we need another ideal point at the other end of $\overline{\ell'}$? As P moves from A towards X, P' moves away from A' towards this ideal point. However, it is also going to be the image of X, so that it should coincide with the X' already chosen. Thus we need only one ideal point per projective line.

© Springer International Publishing AG 2018
A. Liu, *S.M.A.R.T. Circle Minicourses*, Springer Texts in
Education, https://doi.org/10.1007/978-3-319-71743-2_2

It is now clear that the exceptional point Y' on $\overline{\ell'}$ must be the image of the ideal point Y of $\overline{\ell}$.

Note that in Figure 2.1, it appears that B is between A and C on $\overline{\ell}$ while B' is not between A' and C' on $\overline{\ell'}$. Thus betweenness does not seem to be preserved by projections between intersecting lines. This is hardly surprising since betweenness is not even a meaningful concept on a closed curve, and the addition of the ideal point turns a projective line into a closed curve.

Let A, B, C and D be four points on a closed curve. We say that A and B **separate** C and D if in going from A to B along the curve, we must pass over C in one direction and over D in the other. It is easy to see that in that case, C and D also separate A and B, and that separation is preserved by projection.

How would the conversion of parallel Euclidean lines ℓ and ℓ' into projective lines affect projections between them? It is easy to see that there is a one-to-one correspondence between the ordinary points of $\overline{\ell}$ and $\overline{\ell'}$. Let Z be the ideal point of $\overline{\ell}$. The line VZ is supposed to intersect $\overline{\ell'}$ at its ideal point Z'. In order to come up with Z'. we must specify how we are to join an ideal point to an ordinary point. To do this, we must systematically convert the Euclidean plane into the **projective plane**.

In the Euclidean plane, every two points determine a unique line, but not every two lines determine a unique point. The exceptional cases are when the two lines are parallel. If we postulate that parallel lines meet at the same ideal point, we have eliminated this anomaly. Although these lines now meet at an ideal point, we will continue to call them parallel lines.

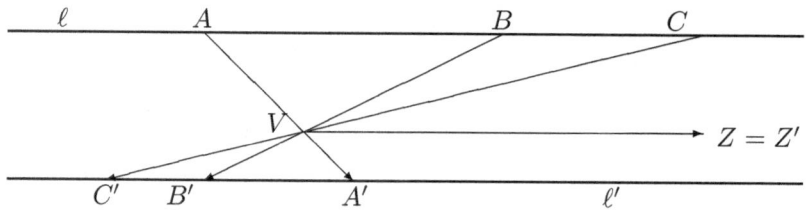

Figure 2.2

We may think of ideal points as the directions of the families of parallel lines. The line VZ is simply the line through V parallel to ℓ. Since it is also parallel to ℓ', they intersect at $Z' = Z$. This agrees with the coincidence of F and F' in Figure 2.1. There, $X' \neq Y$, so that intersecting lines have different ideal points. Otherwise, two such lines will determine more than just one point.

Is it still true that every two points determine a unique line? If both are ordinary points, we have no problem. If one is an ordinary point and the other is an ideal point, they still determine a unique line since we have an ordinary point on it, and we also know its direction. However, two ideal points do not determine any line as yet.

To fix this, we add a new projective line called the **ideal line**, consisting of all ideal points. We now check that every two projective lines still determine a unique point. If both are ordinary, parallel or otherwise, we have no problem. If one is an ordinary line and the other is the ideal line, they determine the ideal point on the ordinary line. Since we have only one ideal line, there are no other cases, and the construction of the projective plane is now completed.

In pure Projective Geometry, there is no distinction between ideal points and ordinary points, or between the ideal line and ordinary lines. Any of the lines may be taken as the ideal line, and the points on it as ideal points. However, our primary interest is to use projective methods to solve Euclidean problems. This is why we retain the notion of parallel lines in the Euclidean sense.

Note that the point V may be an ideal point. In that case, the lines from V are simply parallel lines whose direction is specified by the ideal point V. Projections from an ideal point are often called **parallel projections**, and those from an ordinary point **central projections**. However, this distinction is inconsequential. In many ways, the projective plane is actually simpler than the Euclidean plane.

We can extend Euclidean space into projective space by making each Euclidean plane into a projective plane, and adding an **ideal plane** which contains all ideal lines of the projective planes. In Solid Geometry, a plane becomes a subset of space. Thus it requires an implicit definition, and this is done by listing its basic properties.

1. Two points in a plane lies on a unique line which lies in that plane.

2. Two planes meeting at a point meets along a unique line passing through that point.

3. A line and a point which is not on that line lie on a unique plane.

4. A plane and a line which is not on that plane meet at a unique point.

5. Three points not all on a line lies on a unique plane.

6. Three planes not all meeting along a line meet at a unique point.

7. Two lines on the same plane meet at a unique point.

8. Two lines which meet at a point lie on a unique plane.

Lines and planes that meet at a point are said to be **concurrent**. Points that lie on a line and planes that meet along a line are said to be **collinear**. Points and lines that lie in a plane are said to be **coplanar**. Note that it is possible for two lines not to meet, at either an ordinary or an ideal point. An example is a straight bridge over a straight river. Two such lines are said to be a pair of **skew lines**.

We will retain the Euclidean concept of parallelism. Planes that are parallel meet along their common ideal line. A line and a plane which are parallel meet at the ideal point of the line. A plane is parallel to itself.

Parallel Line-Plane Theorem.
Let AB be a line in a plane Π and CD be a line not in Π. If AB and CD are parallel, then CD is parallel to Π.

Proof:
Suppose to the contrary that CD and Π meet at an ordinary point P. Since AB and CD are parallel, P is not on AB. Moreover, AB and CD lie on a plane. This plane must contain P which is on CD. However, the unique plane containing P and AB is Π, but this contradicts the assumption that CD is not in Π.

While we are primarily interested in Plane Geometry, the embedding of a plane in space gives us new ways of tackling problems. An important tool is the projection of a plane Π to another plane Π' from a point V not on either. There are two basic properties of such a projection. First, a straight line ℓ on Π becomes a straight line ℓ' in Π', ℓ' being the intersection of Π' and the plane determined by V and ℓ. Secondly, a point P in Π lies on ℓ if and only if its image P' in Π' lies on ℓ'.

Let ℓ be any line on a plane be Π. Choose a point V not on Π and let Ω be the unique plane determined by V and ℓ. Let Π' be any plane parallel to but different from Ω. For any point P on ℓ, its image P' is in Ω as well as Π'. Since these two planes are parallel, P' is an ideal point and the image ℓ' of ℓ is the ideal line of Π'. We have projected ℓ to infinity.

Desargues' Theorem.
Let ABC and $A'B'C'$ be two triangles. Suppose AA', BB' and CC' are concurrent at some point V, BC and $B'C'$ meet at L, CA and $C'A'$ meet at M, and AB and $A'B'$ meet at N. Then L M and N are collinear.

Proof:
Project MN to infinity. Then AB and $A'B'$ become parallel lines, as do AC and $A'C'$. Hence VAB and $VA'B'$ are similar triangles, so that $\frac{VB'}{VB} = \frac{VA'}{VA}$. Similarly, $\frac{VC'}{VC} = \frac{VA'}{VA}$. It follows that $\frac{VB'}{VB} = \frac{VC'}{VC}$, so that VBC and $VB'C'$ are also similar triangles. This means that BC and $B'C'$ are parallel, so that L is an ideal point. Hence L, M and N are collinear.

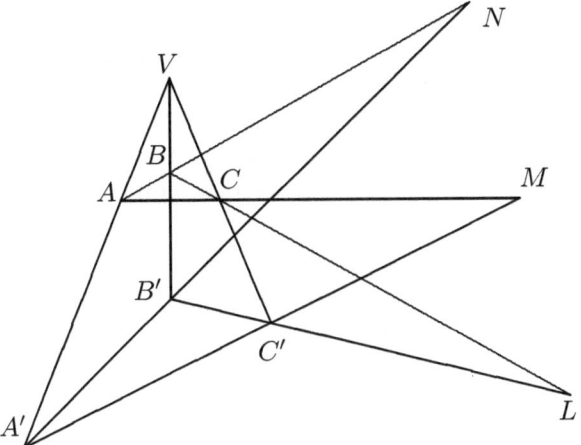

Figure 2.3

Converse of Desargues' Theorem.
Let ABC and $A'B'C'$ be two triangles. Suppose BC and $B'C'$ meet at L, CA and $C'A'$ meet at M, AB and $A'B'$ meet at N, where L, M and N are collinear. Then AA', BB' and CC' are concurrent.

Proof:
Project MN to infinity. Then the corresponding sides of ABC and $A'B'C'$ are parallel, so that the triangles are similar. Let AA' and BB' meet at V. Then VAB and $VA'B'$ are also similar triangles. Thus $\frac{B'C'}{BC} = \frac{A'B'}{AB} = \frac{VB'}{VB}$ while $\angle VB'C' = \angle VBC$. Hence BBC and $VB'C'$ are similar triangles, so that $\angle BVC = \angle B'VC'$. It follows that C, C' and V are collinear, so that AA', BB' and CC' are concurrent.

The Pappus-Pascal Theorem.
A, C and E are collinear points, as are B, D and F. If AB and DE meet at N, BC and EF meet at L, and CD and FA meet at M, then L, M and N are also collinear.

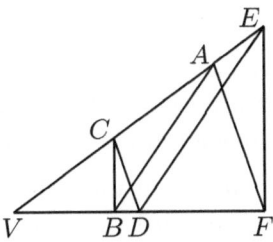

Figure 2.4

Proof:
Project MN to infinity. Then AB and DE are parallel lines, as are CD and FA. Let AE cut BF at V. Then VAB and VED are similar triangles, as are VAF and VCD. Hence $\frac{VA}{VB} = \frac{VE}{VD}$ and $\frac{VA}{VF} = \frac{VC}{VD}$. It follows that $\frac{VC}{VB} = \frac{VE}{VF}$, so that VCB and VEF are similar triangles. This means that BC and EF are parallel, so that L is an ideal point. Hence L, M and N are collinear.

The Pappus-Brianchon Theorem.
AB, CD and EF are concurrent lines, as are BC, DE and FA. Then AD, BE and CF are also concurrent.

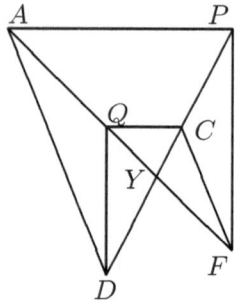

Figure 2.5

Proof:
Let AB, CD and EF be concurrent at P and let BC, DE and FA be concurrent at Q. Let CD cut FA at Y and project BE to infinity. Then AP and QC become parallel lines, as do FP and QD. Hence YAP and YQC are similar triangle, and so are YFP and YQD. It follows that $\frac{YQ}{YA} = \frac{YC}{YP}$ and $\frac{YF}{YQ} = \frac{YP}{YD}$, so that $\frac{YF}{YA} = \frac{YC}{YD}$. Hence YCF and YDA are similar triangles, so that AD and CF are parallel lines. It follows that they are concurrent with the ideal line BE.

An **inaccessible point** is one which is well-defined, usually as the intersection of two lines, but the point itself is not to be drawn.

Here is a construction problem involving an inaccessible point. The point N of intersection of two given lines ℓ and m is inaccessible. Using only a straight-edge, construct the line joining N to a given point M.

Take V not on either ℓ or m. Draw a line through V, cutting ℓ at A and m at D. Draw a second line through V, cutting ℓ at B and m at E. Draw a third line through V, cutting AM at C and DM at F. Let BC cut EF at L. By Desargues' Theorem, the line joining M and N passes through L.

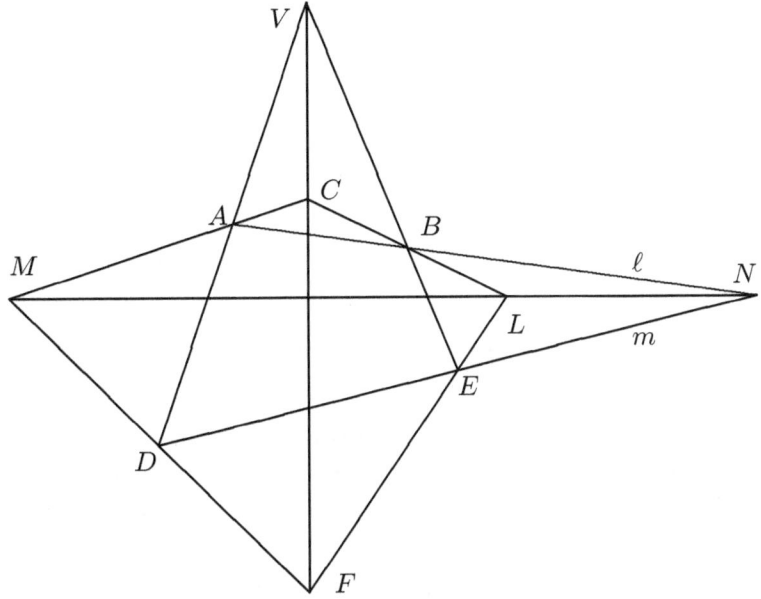

Figure 2.6

An **inaccessible line** is one which is well-defined, usually by two points on it, but the line itself is not to be drawn.

Here is a construction problem involving an inaccessible line. The line ℓ joining two given points A and B is inaccessible. Using only a straight-edge, construct the point of intersection of ℓ with a given line m.

Take V not on either ℓ or m. Let m cut VA at D and VB at E. Draw a third line through V. On this line, take a point C not on ℓ and a point F not on m. Let CA cut FD at M, CB cut FE at L, and LM cut m at N. By Desargues' Theorem, N also lies on ℓ.

Figure 2.6 serves for both constructions. In the former, the point N is to be omitted. In the latter, the line ℓ is to be omitted.

Constructions using a straight edge but without a compass are known as Poncelet-Steiner constructions.

Exercises

1. (a) Let Π and Π' be two planes and V be a point not in either of them. Let triangle ABC in Π be projected from V onto triangle $A'B'C'$ in Π'. Prove that the lines BC and $B'C'$ will intersect, as will CA and $C'A'$, as well as AB and $A'B'$, in three collinear points.

   (b) Deduce Desargues' Theorem from (a).

2. (a) Deduce Desargues' Theorem and its converse from each other.

   (b) Deduce the Pappus-Pascal Theorem and the Pappus-Brianchon Theorem from each other.

3. (a) The point L of intersection of two given lines ℓ and ℓ' is inaccessible, as is the point M of intersection of two given lines m and m'. Using only a straight-edge, construct the line joining L and M.

   (b) The line ℓ joining two given points A and B is inaccessible, as is the line ℓ' joining two given points A' and B'. Using only a straight-edge, construct the point of intersection of ℓ and ℓ'.

Section 2. Metric Approach.

In our synthetic treatment of projective geometry, we have completely avoided the measurement of distances. There are two reasons. First, distances to the ideal elements are necessarily infinite. Secondly, distance is not preserved by projection. However, our goal is to enhance with projective methods the contents of Euclidean Geometry, where distance is an essential concept. In this section, we will find ways to overcome the two difficulties mentioned above.

For a line, we choose arbitrarily either direction as being *positive*, so that the opposite direction is *negative*. Suppose AB is a segment on this line. we denote by $\overline{AB}$ the directed segment going from A to B. Just as the symbol AB also denotes distance, the symbol $\overline{AB}$ also denotes the signed distance. In particular, $\overline{AA} = 0$ and $\overline{AB} + \overline{BA} = 0$.

Let the Cartesian coordinates of A and B be 0 and 1 respectively, so that $\overline{AB} = 1$. For any point P on the line AB, we define the **Apollonian ratio** of P with respect to A and B to be $\dfrac{\overline{AP}}{\overline{PB}}$. Thus if the Cartesian coordinate of P is x, then its Apollonian ratio $\frac{x}{1-x}$. In particular, the Apollonian ratio of A is 0 and that of B is ∞. In Figure 2.7, the Cartesian coordinates are above the line while the Apollonian ratios are below the line.

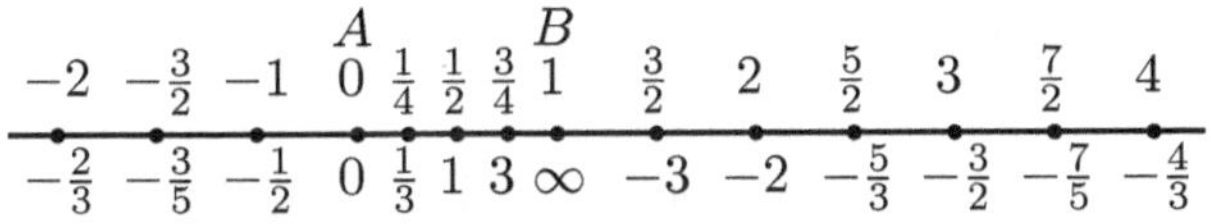

$$
\begin{array}{c}
 \quad \phantom{-\tfrac{3}{2}} \quad \quad A \quad \phantom{\tfrac{1}{4}} \quad \phantom{\tfrac{1}{2}} \quad \phantom{\tfrac{3}{4}} \quad B
\end{array}
$$

$$
\begin{array}{ccccccccccccc}
-2 & -\tfrac{3}{2} & -1 & 0 & \tfrac{1}{4} & \tfrac{1}{2} & \tfrac{3}{4} & 1 & \tfrac{3}{2} & 2 & \tfrac{5}{2} & 3 & \tfrac{7}{2} & 4 \\
-\tfrac{2}{3} & -\tfrac{3}{5} & -\tfrac{1}{2} & 0 & \tfrac{1}{3} & 1 & 3 & \infty & -3 & -2 & -\tfrac{5}{3} & -\tfrac{3}{2} & -\tfrac{7}{5} & -\tfrac{4}{3}
\end{array}
$$

Figure 2.7

If P lies between A and B, both $\overline{AP}$ and $\overline{PB}$ are positive so that its Apollonian ratio is positive. As P moves from A towards B, the numerator increases while the denominator decreases. Hence its Apollonian ratio increases without bound.

Beyond B, $\overline{AP}$ is still positive but $\overline{PB}$ is now negative. Hence the Apollonian ratio of P is negative. As P moves towards B, the absolute value of its Apollonian ratio increases without bound. As P moves away from B, the absolute values of both the numerator and the denominator increase but the former is always 1 greater than the latter. Hence the absolute value of the Apollonian ratio of P decreases to 1 eventually, so that -1 is the natural choice as the Apollonian ratio of the ideal point of the line AB.

On the other side of A, $\overline{AP}$ is negative while $\overline{PB}$ is positive. Hence the Apollonian ratio of P is negative. As P moves towards A, the absolute value of its Apollonian ratio decreases to 0.

As P moves away from A, the absolute values of both the numerator and the denominator increase but the former is always 1 less than the latter. Hence the absolute value of the Apollonian ratio of P increases to 1 eventually. This agrees with the previous choice of -1 as the Apollonian ratio of the ideal point.

Suppose a point P starts from B, moves away from A, reappears at the other end of the line via the ideal point, continues past A and get back to B. During this movement, its Apollonian ratio is always increasing.

It should be pointed out that the choice of the Cartesian coordinates of A and B is immaterial. The Apollonian ratio of the ideal point with respect to any two points on the line will always be -1. Thus the Apollonian ratio is a useful concept in which the ideal point has a finite measure.

Let ABC be any triangle. Let D, E and F be points on the lines BC, CA and AB, respectively, not coinciding with any of A, B and C. The product of the Apollonian ratios $\dfrac{\overline{AF}}{\overline{FB}}$, $\dfrac{\overline{BD}}{\overline{DC}}$ and $\dfrac{\overline{CE}}{\overline{EA}}$ is called a **Cevian ratio**, and is denoted by (ABC, FDE).

We claim that Cevian ratios are preserved by projections. To prove this, we introduce the concept of signed angles. Angles are considered positive if measured in the counterclockwise direction, and negative if measured in the clockwise direction. Like signed distances, signed angles are marked with an overline.

Let V be a point not on the plane of ABC and let d be the distance from V to AB. Then $d\overline{AF} = \pm VA \cdot VF \sin\overline{AVF}$ since the absolute value of each side is twice the area of triangle VAF. Similarly, we also have $d\overline{FB} = \pm VF \cdot VB \sin\overline{FVB}$. Since the right sides of these two equations have the same sign, $\dfrac{\overline{AF}}{\overline{FB}} = \dfrac{VA \sin\overline{AVF}}{VB \sin\overline{FVB}}$. It follows that

$$(ABC, FDE) = \frac{\sin\overline{AVF}\,\sin\overline{BVD}\,\sin\overline{CVE}}{\sin\overline{FVB}\,\sin\overline{DVC}\,\sin\overline{EVA}}.$$

Since angles at V are not affected by projections from V, the claim is justified.

Menelaus' Theorem.
Let ABC be any triangle. Let D, E and F be points on the lines BC, CA and AB, respectively, not coinciding with any of A, B and C. Then D, E and F are collinear if and only if $(ABC, FDE) = -1$.

Proof:

Project EF to infinity. Then $\dfrac{\overline{AF}}{\overline{FB}} = -1 = \dfrac{\overline{CE}}{\overline{EA}}$. Suppose D, E and F are collinear. Then D is an ideal point, and $\dfrac{\overline{BD}}{\overline{DC}} = -1$. It follows that we have $(ABC, FDE) = -1$. Conversely, if $(ABC, FDE) = -1$, then $\dfrac{\overline{BD}}{\overline{DC}} = -1$. It follows that D must be the ideal point of the line BC, and D, E and F are indeed collinear.

Ceva's Theorem.

Let ABC be any triangle. Let D, E and F be points on the lines BC, CA and AB, respectively, not coinciding with any of A, B and C. Then AD, BE and CF are concurrent if and only if $(ABC, FDE) = 1$.

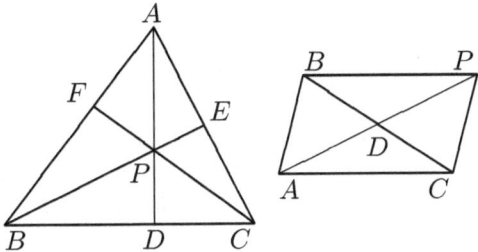

Figure 2.8

Proof:

Let BE and CF meet at P. Project EF to infinity. Then $\dfrac{\overline{AF}}{\overline{FB}} = -1 = \dfrac{\overline{CE}}{\overline{EA}}$. Moreover, $ABPC$ has become a parallelogram. Suppose AD also passes through P. Then D is the point of intersection of the diagonals AP and BC of the parallelogram. Hence $\dfrac{\overline{BD}}{\overline{DC}} = 1$ and $(ABC, FDE) = 1$. Conversely, if $(ABC, FDE) = 1$, then $\dfrac{\overline{BD}}{\overline{DC}} = 1$. It follows that D must be the midpoint of BC. Hence D lies on AP, so that AD, BE and CF are indeed concurrent.

The "only if" parts of these two theorems are often referred to as their respective converses. We now use these theorems to solve a problem.

AD is an altitude of triangle ABC. E and F are points on CA and AB respectively such that BE, CF and AD are concurrent. Prove that AD bisects $\angle FDE$.

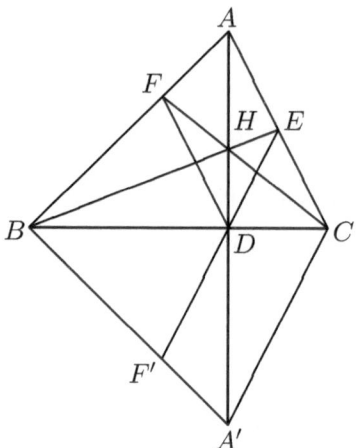

Figure 2.9

Reflect triangle ABC across BC and let A' and F' be the respective images of A and F. Since AD is an altitude, A, D and A' are collinear. Let H be the point of concurrency of AD, BE and CF. Applying Ceva's Theorem to triangle ABH, we have $(ABH, FED) = 1$. It follows that $(A'BH, F'ED) = -1$. Applying the converse of Menelaus' Theorem to triangle $A'BH$, E, D and F' are collinear. Hence

$$\angle ADE = \angle A'DF' = \angle ADF.$$

Note that we can derive Ceva's Theorem from Menelaus' Theorem as follows. Suppose AD, BE and CF are concurrent at P. Applying Menelaus' Theorem to ABD and ADC, we have $(ABD, FCP) = (ADC, PBE) = -1$. Multiplying these together, we have $(ABC, FDE) = 1$. Conversely, let AD cut BE at Q and CF at R. Applying Menelaus's Theorem to ABD and ADC, we have $(ABD, FCQ) = (ADC, RBE) = -1$. Multiplying these together with $1 = (ABC, FDE)$, we have $\dfrac{\overline{AQ}}{\overline{QD}} = \dfrac{\overline{AR}}{\overline{RD}}$. Hence $Q = R$, so that AD, BE and CF are indeed concurrent.

Let A, B, C and D be four distinct collinear points. The **cross ratio** (AB, CD) is defined as the product of the Apollonian ratios $\dfrac{\overline{AC}}{\overline{CB}}$ and $\dfrac{\overline{BD}}{\overline{DA}}$.

In the same way as for Cevian ratios, we can prove that cross ratios are also preserved by projections. Suppose a projection from V sends collinear points A, B, C and D into collinear points A', B', C' and D' respectively. Then $(AB, CD) = (A'B', C'D')$. We incorporate the point V into the equation by writing $(AB, CD)V(A'B', C'D')$. The converse of this result has the following important special case.

Cross Ratio Concurrency Theorem.
Let B, C and D be three points on a line. Let B', C' and D' be three points on another lines. If these lines intersects at A and $(AB, CD) = (AB', C'D')$, then BB', CC' and DD' are concurrent.

Proof:
Let BB' intersect CC' at V. Suppose VD intersects AB' at E'. Then we have $(AB, CD)V(AB', C'E')$, so that $(AB', C'D') = (AB', C'E')$. Hence $\dfrac{\overline{B'D'}}{\overline{D'A}} = \dfrac{\overline{B'E'}}{\overline{E'A}}$, so that $D' = E'$. It follows that VD passes through D', and BB', CC' and DD' are concurrent at V.

We now use this result to prove the theorems in Section 1.

Let MN cut VA, VB and VC at X, Y and Z respectively. Then we have $(VZ, CC')M(VX, AA')N(VY, BB')$. By the Cross Ratio Concurrency Theorem, $BC, B'C'$ and XY are concurrent. It follows that L, M and N are collinear. This establishes Desaegues' Theorem.

Let AA' cut BC, $B'C'$ and MN at X, X' and Y respectively. Then we have $(LB, XC)A(LN, YM)A'(LB', X'C')$. By the Cross Ratio Concurrency Theorem, AA', BB' and CC' are concurrent. This establishes the converse of Desargues' Theorem.

Let P be the point of concurrency of AB, CD and EF, and Q be that of BC, DE and FA. Let AB cut CF at X, BC cut EF at Y, CD cut FA at Z, and BE cut CF at R. Then $(RX, FC)B(EP, FY)Q(DP, ZC)$. By the Cross Ratio Concurrency Theorem, RD, XP and FZ are concurrent. It follows that AD, BE and CF are also concurrent. This establishes the Pappus-Pascal Theorem.

Let CD cut EF at X, DE cut FA at Y and AC cut BD at Z. Then $(XL, EF)C(DB, ZF)A(DN, EY)$. By the Cross Ratio Concurrency Theorem, XD, LN and FY are concurrent. It follows that L, M and N are collinear. This establishes the Pappus-Brianchon Theorem.

Cross Ratio Equality Theorem.
$(AB, CD) = (DC, BA) = (CD, AB) = (BA, DC)$ for any four collinear points A, B, C and D.

Proof:
Take a point V not on AD and join it to A, B and C. Draw a line through D, cutting VC at E, VB at F and VA at G. Let GC cut VB at M. Then $(AB, CD)G(VB, MF)C(ED, GF)V(CD, AB)$. Let FA cut VC at N. Then $(AB, CD)F(NV, CE)A(FG, DE)V(BA, DC)$. Let FC cut VA at L. Then $(AB, CD)F(AV, LG)C(DE, FG)V(DC, BA)$.

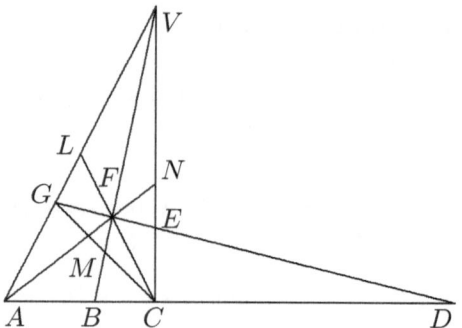

Figure 2.10

There are 24 permutations of the points A, B, C and D, so that they give rise to 24 cross ratios. In general, they form six groups of four and take on distinct values.

We tend not to make use of the actual value of cross ratios. An exception is the special case where $(AB, CD) = -1$. We say that C are D are **harmonic conjugates** of each other respect to A and B, and the four points A, C, B and D form a **harmonic range**. The harmonic conjugate of the midpoint of AB with respect to A and B is the ideal point of AB.

Harmonic Range Theorem.
If A, C, B and D form a harmonic range, then $(AB, CD) = (AB, DC)$.

Proof:
Given A, B and C, we first give a Poncelet-Steiner construction of D. Take a point V not on AB and a point E on VC. Let EA cut VB at F and EB cut VA at G. The point D on intersection of FG and AB is the desired conjugate. By Ceva's Theorem, $(VAB, GCF) = 1$. By Menelaus' Theorem, $(VBA, FDG) = -1$. It follows that $(AB, CD) = -1$. Let VC cut FG at H. It follows from the Cross Ratio Equality Theorem that we have $(AB, CD) = (BA, DC)E(GF, DH)V(AB, DC)$.

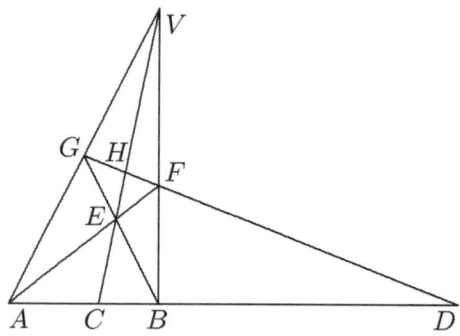

Figure 2.11

Exercises:

4. Deduce from Menelaus' Theorem

   (a) Desargues' Theorem;

   (b) the converse of Desargues' Theorem;

   (c) the Pappus-Pascal Theorem;

   (d) the Pappus-Brianchon Theorem.

5. (a) The bisector of the exterior angle at the vertex A of triangle ABC meets the extension of BC at K while the bisector of $\angle CAB$ meets BC at D. A line through K cuts CA at E and AB at F. Prove that AD, BE and CF are concurrent.

   (b) The bisector of the exterior angle at the vertex A of triangle ABC meets the extension of BC at K while the bisector of $\angle CAB$ meets BC at D. P is a point on AD. The extensions of BP and CP cut CA and AB at E and F, respectively. Prove that E, F and K are collinear.

   (c) D, E and F are points on the sides BC, CA and AB of triangle ABC respectively. The lines BC and EF meet at L, the lines CA and FD meet at M, and the lines AB and DE meet at N. If L, M and N are collinear, prove that AD, BE and CF are concurrent.

6. (a) Determine (DC, BA) in terms of (AB, CD).

   (b) Determine (AB, DC) in terms of (AB, CD).

   (c) P is a point inside triangle ABC. AP cuts BC at D, BP cuts CA at E, CP cuts AB at F, EF cuts BC at X, FD cuts CA at Y and DE cuts AB at Z. Prove that X, Y and Z are collinear.

Section 3. Analytic Approach.

We begin with some analytic geometry but in a rather unusual fashion. Consider a typical line $3x + 2y - 6 = 0$ which does not pass through the origin. It is a collection of points (x, y) where $y = 2 - \frac{2}{3}x$. We can plot this line by computing points on it.

x	-2	-1	0	1	2	3	4	5
y	$\frac{10}{3}$	$\frac{8}{3}$	2	$\frac{4}{3}$	$\frac{2}{3}$	0	$-\frac{2}{3}$	$-\frac{4}{3}$

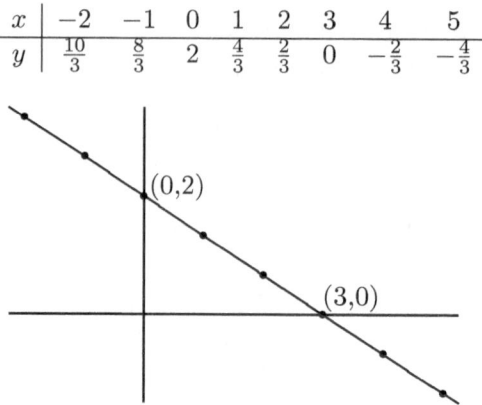

Figure 2.12

Since the constant term is non-zero, we divide the equation by it so that it now has the form $-\frac{1}{3}x - \frac{1}{2}y + 1 = 0$. The general form is $ux + vy + 1 = 0$. Here $u = -\frac{1}{3}$ and $v = -\frac{1}{2}$. Note that $-\frac{1}{u} = 3$ is the x-intercept of the line while $-\frac{1}{v} = 2$ is the y-intercept of the line.

We have seen in Section 1 the symmetry between points and lines in the projective plane. If we can give Cartesian coordinates to points, we should be able to give some kind of coordinates to lines too. From the symmetry in the equation $ux + vy + 1 = 0$ between x and y on the one hand versus u and v on the other hand, it is reasonable to define its **line coordinates** to be $[u, v]$. We use square brackets to remind ourselves that we are not talking about point coordinates.

In the equation $-\frac{1}{3}x - \frac{1}{2}y + 1 = 0$, we have treated x and y as variables and give specific numerical values to u and v. What we get is a line. What will we get if we treat u and v as variables and give specific numerical values to x and y? For instance, what is represented by $3u - 2v + 1 = 0$?

Just as $-\frac{1}{3}x - \frac{1}{2}y + 1 = 0$ represent all the points on the line $2x + 3y - 6 = 0$, we expect $3u - 2v + 1 = 0$ to represent all the lines passing through some point. Clearly, this point should be $(3, -2)$. Let us verify this graphically.

u	-1	1	$\frac{1}{3}$	$\frac{1}{5}$
v	-1	2	1	$\frac{4}{5}$

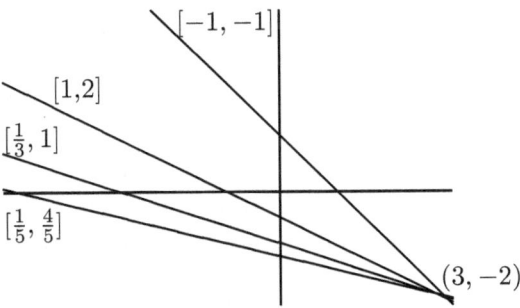

Figure 2.13

Consider the horizontal line passing through $(3, -2)$. The x-intercept approaches infinity, so that u approaches 0. Setting $u = 0$ in $3u - 2v + 1 = 0$, we have $v = \frac{1}{2}$. Hence the desired line coordinates are $[0, \frac{1}{2}]$. Note that its y-intercept is indeed $-\frac{1}{v} = -2$. Similarly, the line coordinates of the vertical line passing through $(3, -2)$ is $[-\frac{1}{3}, 0]$.

In general, $[u, 0]$ is the vertical line with x-intercept $-\frac{1}{u}$ while $[0, v]$ is the horizontal line with y-intercept $-\frac{1}{v}$. Note that $[0,0]$ does not represent any line. So now every line not passing through the origin has line coordinates.

The equation in point coordinates of the unit circle is $x^2 + y^2 = 1$, representing all points which lie on the circle. What should the equation be in line coordinates? What are the lines which should satisfy this equation? Lines can define a circle just as well as points. If we draw all lines tangent to the unit circle, they will form what is called an *envelope* of the circle.

Solving $x^2 + y^2 = 1$ for x, we have $x = \pm\sqrt{1 - y^2}$. Substituting this into $ux + vy + 1 = 0$, we have $vy + 1 = \mp u\sqrt{1 - y^2}$. Squaring both sides, we have $v^2y^2 + 2vy + 1 = u^2(1 - y^2)$. This may be rewritten as a quadratic equation in y, namely, $(u^2 + v^2)y^2 + 2vy + (1 - u^2) = 0$. If it has two real roots, they are the y coordinates of the points of intersection of the unit circle with this line. Since we want this line to be a tangent, the discriminant of the equation should be zero. Now $4v^2 - 4(u^2 + v^2)(1 - u^2) = 0$ simplifies to $u^2(u^2 + v^2 - 1) = 0$. Apart from the singularities represented by the two horizontal tangents, $u \neq 0$, so that the equation of the unit circle in line coordinates is $u^2 + v^2 = 1$, another triumph of the symmetry between points and lines.

However, as in Section 1, this symmetry is incomplete if we confine ourselves to the Euclidean plane. Lines passing through the origin do not have line coordinates. To achieve complete symmetry, we must once again extend to the projective plane, but then we need some way of coordinatizing ideal elements. This can be done by means of **homogeneous coordinates**.

In the Euclidean plane, a line is represented by an equation of the form $Ax + By + C = 0$, which equation is non-homogeneous if $C \neq 0$. This is because two of the terms are of degree 1 while the third term is of degree 0.

We now coordinatize a point (x, y) as $(\overline{x}, \overline{y}, \overline{z})$ where $\overline{z}$ is any non-zero number such that $\frac{\overline{x}}{\overline{z}} = x$ and $\frac{\overline{y}}{\overline{z}} = y$. Just as both $\frac{3}{5}$ and $\frac{6}{10}$ represent the same fraction, both $(3,5,1)$ and $(6,10,2)$ represent the same point, namely $(3,5)$. In general, $(\overline{\lambda x}, \overline{\lambda y}, \overline{\lambda z})$ represent the same point for any non-zero real number λ.

At first, it may seem that we are moving in the wrong direction, in that things become somewhat more complicated than before. If we insist that $\overline{z}$ be non-zero, then this is indeed much ado for nothing. However, by letting $\overline{z}$ tend to 0, both x and y tend to infinity provided $\overline{x}$ and $\overline{y}$ are non-zero. We have found a way to coordinatize ideal points, by setting $\overline{z} = 0$.

Note also that $Ax + By + C = 0$ becomes $A\overline{x} + B\overline{y} + C\overline{z} = 0$, which is always a homogeneous equation. This is the reason for the name homogeneous coordinates. As we shall see, there are significant advantages working with a homogeneous equation over working with a non-homogeneous one.

We first establish homogeneous point coordinates. We have no use for the triple $(0,0,0)$. Every other triple represents some point in the projective plane. As we have seen, if $\overline{z} \neq 0$, $(\overline{x}, \overline{y}, \overline{z})$ represent an ordinary point (x, y) where $x = \frac{\overline{x}}{\overline{z}}$ and $y = \frac{\overline{y}}{\overline{z}}$.

The ideal point on a line with slope $\frac{y}{x}$ is represented by $(\overline{x}, \overline{y}, 0)$. The ideal point on the x-axis is represented by $(\overline{x}, 0, 0)$ and the ideal point on the y-axis is represented by $(0, \overline{y}, 0)$. The last two points are more conveniently represented by $(1,0,0)$ and $(0,1,0)$ respectively.

We now turn our attention to homogeneous line coordinates. In particular, we have to take care of lines passing through the origin, as well as the ideal line. It is perhaps not surprising that $[u, v]$ now becomes $[\overline{u}, \overline{v}, \overline{w}]$, where $\frac{\overline{u}}{\overline{w}} = u$ and $\frac{\overline{v}}{\overline{w}} = v$. Once again, we have no use for the triple $[0,0,0]$, and $[\overline{\lambda u}, \overline{\lambda v}, \overline{\lambda w}]$ represent the same line for any non-zero real number λ.

If $\overline{w} \neq 0$, $[\overline{u}, \overline{v}, \overline{w}]$ is the ordinary line $[u, v]$ which does not pass through the origin. The homogeneous coordinates for the ideal points suggest that the homogeneous equation for the ideal line should be $\overline{w}\overline{z} = 0$, so that its line coordinates may be taken as $[0,0,1]$. Thus we have found a use for the pair $[0,0]$ which was left out before.

The homogeneous equation for a line passing through the origin has the form $\overline{ux} + \overline{yv} = 0$ so that its line coordinates are $[\overline{u}, \overline{v}, 0]$. In particular, the line with slope $\frac{3}{5}$ has line coordinates $[3, -5, 0]$. The line coordinates for the x-axis is $[0,1,0]$ and those for the y-axis are $[1,0,0]$.

We are now all set to do some analytic geometry in homogeneous coordinates. Note that the equation $\overline{u}x + \overline{v}y + \overline{w}z = 0$ holds if and only if the point $(\overline{x}, \overline{y}, \overline{z})$ lies on the line $[\overline{u}, \overline{v}, \overline{w}]$. The expression $\overline{u}x + \overline{v}y + \overline{w}z$ is called the *dot product* of $(\overline{x}, \overline{y}, \overline{z})$ and $[\overline{u}, \overline{v}, \overline{w}]$, the sum of the products of corresponding terms. Naturally, it is denoted by a dot, and is a commutative operation.

Our first task is to find the homogeneous coordinates $(\overline{x}, \overline{y}, \overline{z})$ of the point of intersection of two lines whose homogeneous coordinates are $[\overline{u_1}, \overline{v_1}, \overline{w_1}]$ and $[\overline{u_2}, \overline{v_2}, \overline{w_2}]$. That these are lines means that neither triple is $[0,0,0]$. That they are distinct lines means that $\frac{\overline{u_1}}{\overline{u_2}} = \frac{\overline{v_1}}{\overline{v_2}} = \frac{\overline{w_1}}{\overline{w_2}}$ does not hold. By symmetry, we may assume that $\overline{u_1 v_2} \neq \overline{v_1 u_2}$.

The incidence relations yield a system of simultaneous equations, namely, $\overline{x u_1} + \overline{y v_1} + \overline{z w_1} = 0$ and $\overline{x u_2} + \overline{y v_2} + \overline{z w_2} = 0$. Solving for $\overline{x}$ and $\overline{y}$ in terms of $\overline{z}$, we have $\overline{x} = \frac{\overline{v_1 w_2} - \overline{w_1 v_2}}{\overline{u_1 v_2} - \overline{v_1 u_2}}\overline{z}$ and $\overline{y} = \frac{\overline{w_1 u_2} - \overline{u_1 w_2}}{\overline{u_1 v_2} - \overline{v_1 u_2}}\overline{z}$. It follows that we have $(\overline{x}, \overline{y}, \overline{z}) = (\overline{v_1 w_1} - \overline{w_1 v_2}, \overline{w_1 u_2} - \overline{u_1 w_2}, \overline{u_1 v_2} - \overline{v_1 u_2})$.

This triple is called the *cross product* of $[\overline{u_1}, \overline{v_1}, \overline{w_1}]$ and $[\overline{u_2}, \overline{v_2}, \overline{w_2}]$. Naturally, it is denoted by a cross. It is anti-commutative in that if we reverse the order of the two factors, the terms in the product change signs but retain the same numerical values. Thus in homogeneous coordinates, it does not matter. Figure 2.14 shows a schematic computation of the cross product.

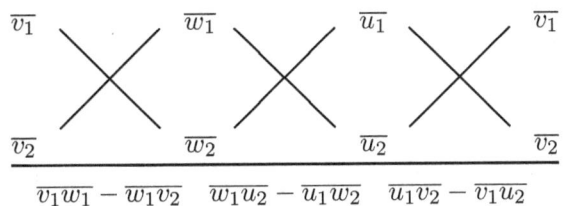

$$\overline{v_1 w_1} - \overline{w_1 v_2} \qquad \overline{w_1 u_2} - \overline{u_1 w_2} \qquad \overline{u_1 v_2} - \overline{v_1 u_2}$$

Figure 2.14

The companion task of finding the homogeneous coordinates $[\overline{u}, \overline{v}, \overline{w}]$ of the line passing through the two points whose homogeneous coordinates are $(\overline{x_1}, \overline{y_1}, \overline{z_1})$ and $(\overline{x_2}, \overline{y_2}, \overline{z_2}]$ parallels the above task, with the cross product of these two triples as the end result, on the assumption that $\overline{x_1 y_2} \neq \overline{y_1 x_2}$.

Let us use homogeneous coordinates to prove Desargues' Theorem. In Figure 2.3, choose V to be $(1,1,1)$, A to be $(1,0,0)$, B to be $(0,1,0)$ and C to be $(0,0,1)$. Then the line VA is $(1, 1, 1) \times (1, 0, 0) = [0, 1, -1]$, which is $\overline{y} = \overline{z}$. Now A' is another point on this line. Hence its coordinates are $(a, 1, 1)$, with $a \neq 1$ to make is distinct from A. Similarly, B' is $(1, b, 1)$ and C' is $(1, 1, c)$, with $b \neq 1 \neq c$.

There is one other condition, that these three points are not collinear. Now the line $A'B'$ is $(a, 1, 1) \times (1, b, 1) = [1 - b, 1 - a, ab - 1]$. If C' lies on it, then $(1, 1, 1) \cdot [1 - b, 1 - a, ab - 1] = 0$. It follows that we must have $(1 - b) + (1 - a) + c(ab - 1) = abc + 2 - a - b - c \neq 0$.

The line AB is $(1,0,0) \times (0,1,0) = [0,0,1]$. The point N is therefore $[0,0,1] \times [1-b, 1-a, ab-1] = (a-1, 1-b, 0)$. Similarly, the point M is $(a-1, 0, 1-c)$ and the point L is $(0, b-1, 1-c)$. The line MN is $(a-1, 0, 1-c) \times (a-1, 1-b, 0) = [(b-1)(1-c), (a-1)(1-c), -(a-1)(b-1)]$. Since $(0, b-1, 1-c) \cdot [(b-1)(1-c), (a-1)(1-c), -(a-1)(b-1)] = 0$, L indeed lies on MN. Hence L, M and N are collinear, proving Desargues' Theorem.

We point out that the coordinates $(1,1,1)$, $(1,0,0)$, $(0,1,0)$ and $(0,0,1)$ can be assigned arbitrarily to any four points provided no three of them are collinear.

We now use homogeneous coordinates to prove the Pappus-Pascal Theorem. In Figure 2.15, choose E to be $(1,1,1)$, C to be $(1,0,0)$, F to be $(0,1,0)$ and N to be $(0,0,1)$. Then the line CE is $(1,1,1) \times (1,0,0) = [0,1,-1]$, and as before, A is $(a,1,1)$ for some $a \neq 1$. Similarly, L is $(1, \ell, 1)$ and D is $(1,1,d)$ for some $\ell \neq 1 \neq d$.

The line AN is $(a,1,1) \times (0,0,1) = [1,-a,0]$. On the other hand, the line CL is $(1,0,0) \times (1,\ell,1) = [0,-1,\ell]$. It follows that the point B is $[1,-a,0] \times [0,-1,\ell] = (-a\ell, -\ell, -1)$. Since this point lies on DF, which is $(0,1,0) \times (1,1,d) = [-d,0,1]$, we must have $(-a\ell, -\ell, -1) \cdot [-d,0,1] = 0$. This simplifies to $\ell a d - 1 = 0$.

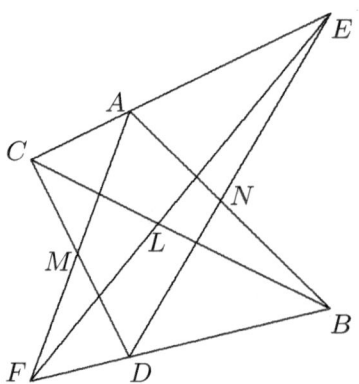

Figure 2.15

The line AF is $(a,1,1) \times (0,1,0) = [-1,0,a]$. On the other hand, the line CD is $(1,0,0) \times (1,1,d) = [0,-d,1]$. It follows that the point M is $[-1,0,a] \times [0,-d,1] = (ad,1,d)$. Now LN is $(1,\ell,1) \times (0,0,1) = [\ell,-1,0]$. Since $(ad,1,d) \cdot [\ell,-1,0] = \ell a d - 1 = 0$, M indeed lies on LN. Hence L, M and N are collinear, proving the Pappus-Pascal Theorem.

This theorem has a surprising algebraic consequence. In Figure 2.4, let $VC = VE = x$, $VA = VY = y$ and $VB = 1$. Now $\frac{VE}{VD} = \frac{VA}{VB}$ or $VE = yx$. On the other hand, $\frac{VE}{VF} = \frac{VC}{VB}$ or $VE = xy$. Thus the Pappus-Pascal Theorem implies that $xy = yx$, the commutative law of multiplication!

Exercises:

7. The equation $2y^2 = x$ in point coordinates represents a parabola, a curve which we will study in the next chapter. Find its equation in line coordinates.

8. Use homogeneous coordinates to prove the converse of Desargues' Theorem.

9. Use homogeneous coordinates to prove the Pappus-Brianchon Theorem.

Bibliography

[1] B. I. Argunov and L. A. Skornyakov, *Configuration Theorems*, D. C. Heath, Boston, 1963.

[2] H. Dorwart, *The Geometry of Incidence*, Prentice Hall, New York, 1966.

[3] H. Eves, *A Survey of Geometry*. 2 vols. Allyn and Bacon, Boston, 2nd edition, 1972.

[4] D. Hilbert and S. Cohn-Vossen, *Geometry and the Imagination*, AMS, Providence, 1983.

[5] C. S. Ogilvy, *Excursions in Geometry*, Dover, Mineola, 1969.

Solution to Exercises

1. (a) Note that B' and C' lie on the plane VBB', so that the coplanar lines BC and $B'C'$ will intersect at some point L. Similarly, CA cuts $C'A'$ at some point M, and AB cuts $A'B'$ at some point N. Now Π and Π' meet along a line, as shown in Figure 2.3. Since L lies in both Π and Π', it lies on this line. Similarly, so do M and N. Hence these three points are collinear.

   (b) Through V, draw a line not on the plane Π of ABC. Take on this line points D and D'. Now BCD and $B'C'D'$ are two non-coplanar triangles satisfying the hypothesis of Desargues' Theorem in Space. Hence BD meets $B'D'$ at some point Y, and CD meets $C'D'$ at some point Z. Moreover, L, Y and Z are collinear. Similarly, AD meets $A'D'$ at some point X, and M, Z and X are collinear, as are N, X and Y. Note that none of X, Y and Z is in Π. A plane Ω containing X, Y and Z will also contain L, M and N, and L, M and N all lie on the line of intersection of Π and Ω.

2. (a) We first deduce the converse from Desargues' Theorem. Consider triangles $BB'N$ and $CC'M$. Note that $B'N$ cuts $C'M$ at D while NB cuts MC at A. Since BC cuts $B'C'$ at L, the lines BB' and CC' are coplanar, and will meet at some point V. Since BC, $B'C'$ and MN are concurrent at L, it follows from Desargues' Theorem that V, A and A' are collinear, which is equivalent to the desired result.

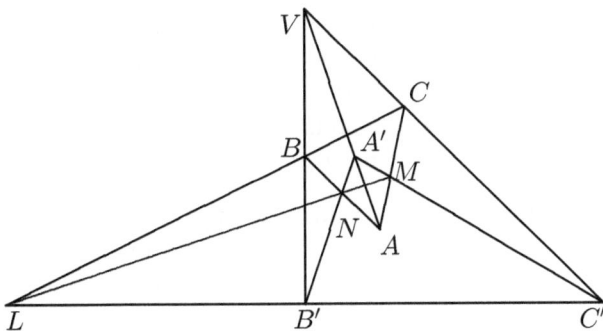

Figure 2.16

We now deduce Desargues' Theorem from its converse. Consider triangles $BB'N$ and $CC'M$. BB' cuts CC' at V, $B'N$ cuts $C'M$ at A', and NB cuts MC at A. Since V, A' and A are collinear, $BC, B'C'$ and MN are concurrent by the converse of Desargues' Theorem. Since BC and $B'C'$ meet at L, L, M and N are collinear.

(b) We first deduce the Pappus-Brianchon Theorem from the Pappus-Pascal Theorem. Let the two given points of concurrency be P and Q respectively, and let BE and CF meet at R. Now E, F and P are collinear, as are B, C and Q. A is the point of intersection of BP and FQ, and D is the point of intersection of PC and QE. It follows from the Pappus-Pascal Theorem that A, D and R are collinear. In other words, AD is concurrent with BE and CF.

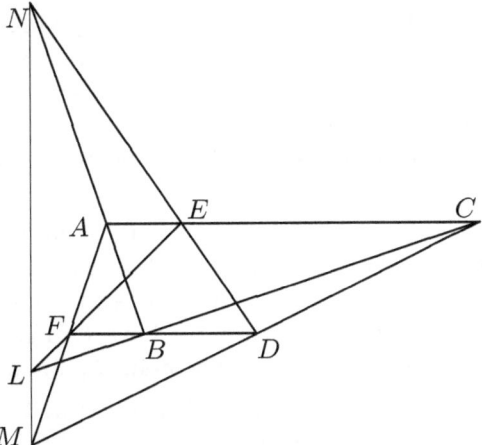

Figure 2.17

We now deduce the Pappus-Pascal Theorem from the Pappus-Brianchon Theorem. Note that AE, LB and DM are concurrent at C while EL, BD and MA are concurrent at F. By the Pappus-Brianchon Theorem, AB, ED and ML are also concurrent. Since AB and DE meet at N, L, M and N are collinear.

3. (a) Let C be the point of intersection of ℓ and m, and C' be that of ℓ' and m'. Take a third point V on CC'. Draw a line through V, cutting ℓ at A and ℓ' at A'. Draw a second line through V, cutting m at B and m' at B'. Let AB cut $A'B'$ at N. Apply Desargues' Theorem to triangles ABC and $A'B'C'$. Since AA', BB' and CC' are concurrent at V, L, M and N are collinear. Repeat this construction varying the position of V on CC'. The line joining the old and new N is the desired line LM.

   (b) Let AA' cut BB' at V and draw a third line m through V. Take C and C' on m. Let AC cut $A'C'$ at M and BC cut $B'C'$ at L. Apply the converse of Desargues' Theorem to triangles MAA' and LBB'. Since C, C' and V ar collinear, LM is concurrent with ℓ and ℓ'. Repeat this construction varying the positions of C and C' on m. The point of intersection of the old and the new LM is the desired point of intersection of ℓ and ℓ'.

4. (a) Applying Menelaus' Theorem to BCV, CAV and ABV, we have $(BCV, LC'B') = (CAV, MA'C') = (ABV, NB'A') = -1$. Multiplying these together, we have $(ABC, NLM) = -1$. By the converse of Menelaus' Theorem, L, M and N are collinear.

   (b) Let AA' and BB' meet at V. Applying Menelaus' Theorem to NBB', NBL and $NB'L$, we have $(NBB', AVA') = -1$ and $(NBL, ACM) = (NB'L, A'C'M) = -1$. Multiplying these together, we have $(LBB', CVC') = -1$. C, C' and V are collinear by the converse of Menelaus' Theorem, so that AA', BB' and CC' are concurrent.

   (c) Let BC cut DE at X, DE cut FA at Y, and FA cut BC at Z. By Menelaus' Theorem, $(XYZ, EFL) = (XYZ, DMC) = -1$ and $(XYZ, NAB) = (XZY, CAE) = (XZY, BFD) = -1$. Multiplying these together, we have $(XYZ, NML) = -1$. By the converse of Menelaus' Theorem, L, M and N are collinear.

   (d) Let BC cut EF at X, DE cut FA at Y, EF cut AB at Z, and AD cut CF at R. We have $(ADY, RCF) = (DQY, EFP) = -1$ as well as $(BCP, QYA) = (PCB, DQZ) = (QZA, EPF) = -1$ by Menelaus' Theorem. Multiplying these together, we then have $(ADZ, REB) = -1$. By the converse of Menelaus' Theorem, B, E and R are collinear, so that AD, BE and CF are concurrent at R.

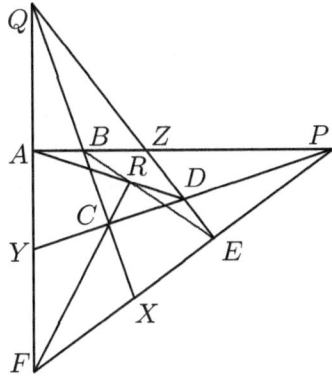

Figure 2.18

5. (a) In ABC, Menelaus' Theorem implies $(ABC, FKE) = -1$. Since BK and BD are angle bisectors, $-\frac{\overline{BK}}{\overline{KC}} = \frac{AB}{AC} = \frac{\overline{BD}}{\overline{DC}}$. Hence $(ABC, FDE) = 1$. By the converse of Ceva's Theorem, AD, BE and CF are concurrent.

   (b) Applying Ceva's Theorem to ABC, we have $(ABC, FDE) = 1$. Since BD and BK are angle bisectors, $\frac{\overline{BD}}{\overline{DC}} = \frac{AB}{AC} = -\frac{\overline{BK}}{\overline{KC}}$. Hence $(ABC, FKE) = -1$. By the converse of Menelaus' Theorem, K, E and F are collinear.

   (c) Menelaus' Theorem implies $(ABC, FLE) = (ABC, FDM) = -1$ and $(ABC, NDE) = -1$. Multiplying these together, we have $(ABC, NLM)(ABC, FDE)^2 = -1$. Since L, M and N are collinear, $(ABC, NLM) = -1$. Hence $(ABC, FDE) = \pm 1$. Since D, E and F all lie on the perimeter of ABC, they cannot be collinear. Hence $(ABC, FDE) \neq -1$. By the converse of Ceva' Theorem, AD, BE and CF are concurrent.

6. (a) We have $(DC, BA) = \frac{\overline{DB \cdot AC}}{\overline{BC \cdot DA}} = \frac{\overline{DB \cdot AC}}{\overline{CB \cdot AD}} = (AB, CD)$.

   (b) We have $(AB, DC) = \frac{\overline{AD \cdot CB}}{\overline{DB \cdot AC}} = \frac{1}{(AB, CD)}$.

   (c) We have $(BD, CX)P(EQ, FX)$ where Q is the point of intersection of EF and AD. Now $(EQ, FX)D(EA, YC)$. By (a) and (b), $(EA, YC) = \frac{1}{(EA, CY)} = \frac{1}{(YC, AE)} = (YC, EA) = (AE, CY)$. From $(BD, CX) = (AE, CY)$, AB, DE and XY are concurrent. Since AB cuts DE at Z, X, Y and Z are collinear.

7. We have $2uy^2 + vy + 1 = 0$, a quadratic equation with discriminant $v^-4(2u) = 0$. It follows that the desired equation in line coordinates is $v^2 - 8u = 0$.

8. In Figure 2.16, omit the line VA. Choose LN to be $[1,1,1]$, BC to be $[1,0,0]$, CA to be $[0,1,0]$ and AB to be $[0,0,1]$. Then the point L is $[1,1,1] \times [1,0,0] = (0,1,-1)$. Now $B'C'$ is another line passing through N. Hence its coordinates are $[a,1,1]$ for some $a \neq 1$. Similarly, $C'A'$ is $[1,b,1]$ and $A'B'$ is $[1,1,c]$, with $b \neq 1 \neq c$. Note that C' is $[a,1,1] \times [1,b,1] = (1-b,1-a,ab-1)$. On the other hand, C is $[1,0,0] \times [0,1,0] = (0,0,1)$. It follows that the line CC' is given by $(0,0,1) \times (1-b,1-a,ab-1) = [a-1,1-b,0]$. Similarly, the line BB' is $[a-1,0,1-c]$ and the line AA' is $[0,b-1,1-c]$. Let V be the point of intersection of BB' and CC'. Then its homogeneous coordinates are $[a-1,0,1-c] \times [a-1,1-b,0] = ((b-1)(1-c),(a-1)(1-c),-(a-1)(b-1))$. Since $[0,b-1,1-c] \cdot ((b-1)(1-c),(a-1)(1-c),-(a-1)(b-1)) = 0$, AA' indeed passes through V. Hence AA', BB' and CC' are concurrent, proving the converse of Desargues' Theorem.

9. In Figure 2.18, omit the line CF and the points X, Y and Z. Choose R to be $(1,1,1)$, A to be $(1,0,0)$, C to be $(0,1,0)$ and E to be $(0,0,1)$. Then AF is $(1,0,0) \times (1,1,1) = [0,-1,1]$. Since Q is another point on this line, take Q to be $(q,1,1)$ for some $q \neq 1$. Similarly, P may be taken as $(1,1,p)$ for some $p \neq 1$. We perform the following computations.

QC	:	$(0,0,1)$	$\times$	$(q,1,1)$	$=$	$[1,0,-q]$
PC	:	$(0,1,0)$	$\times$	$(1,1,p)$	$=$	$[p,0,-1]$
PA	:	$(1,0,0)$	$\times$	$(1,1,p)$	$=$	$[0,-p,1]$
QE	:	$(0,0,1)$	$\times$	$(q,1,1)$	$=$	$[-1,q,0]$
B	:	$[0,-p,1]$	$\times$	$[1,0,-q]$	$=$	$(pq,1,p)$
D	:	$[-1,q,0]$	$\times$	$[p,0,-1]$	$=$	$(-q,-1,-pq)$
AD	:	$(1,0,0)$	$\times$	$(-q,-1,-pq)$	$=$	$[0,pq,-1]$
BE	:	$(0,0,1)$	$\times$	$(pq,1,p)$	$=$	$[-1,pq,0]$
R	:	$[0,pq,-1]$	$\times$	$[-1,pq,0]$	$=$	(pq,q,pq)
CF	:	$(0,1,0)$	$\times$	$(1,1,1)$	$=$	$(1,0,-1)$

Since $(1,0,-1) \cdot [pq,0,pq] = 0$, R indeed lies of CF. Hence AD, BE and CF are concurrent, proving the Pappus-Brianchon Theorem.

Chapter Three: Conic Sections

Section 1. Loci.

A vertical wall and a horizontal floor meet at the point X. A vertical ladder AB is standing against the wall. A cat jumps onto it and sits at its midpoint C. This causes the ladder to move, with its apex A sliding along the wall down towards the floor, and its bottom B sliding along the floor away from the wall. The whole motion takes place in a vertical plane, and comes to a stop when the ladder is lying horizontally on the floor. What is the curve traced by the cat?

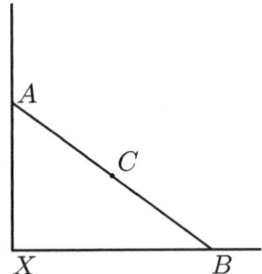

Figure 3.1

Such problems are called *loci* problems. The set of all possible positions of a point satisfying certain conditions is called the locus of the point. For instance, the locus of a point at a fixed distance from a fixed point is a circle, while the locus of a point at a fixed distance from a fixed line is a pair of lines, one on each side of the given line, and parallel to it.

In the problem with the cat on the sliding ladder, complete the rectangle $AXBY$. Then C is also the midpoint of the other diagonal XY. Although A and B are variable points, the length AB is fixed, and so is the length XY. Since CX is half this length, C is at a fixed distance from the fixed point X. Hence its locus is a quarter of a circle.

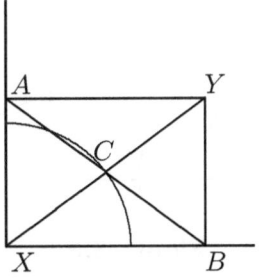

Figure 3.2

© Springer International Publishing AG 2018
A. Liu, *S.M.A.R.T. Circle Minicourses*, Springer Texts in
Education, https://doi.org/10.1007/978-3-319-71743-2_3

Suppose the cat sits at a point C on AB other than its midpoint. If the ladder moves in the same way as before, what is the curve traced by the cat then? To answer this question, we need to know more loci. One of the best book on this subject is [5], from which the above problem comes. We shall return to this problem in Section 3.

The locus of a point P equidistant from two fixed points F and D is the perpendicular bisector of FD, and it intersects the line FD at the midpoint of the segment FD. In the projective plane, there is a second point of intersection, namely, the ideal point of the line FD, and the locus has another branch, namely, the ideal line of the projective plane.

The distance between a point P and a line d, denoted by Pd, is the length of the segment PD where D is the point on d such that PD is perpendicular to d. If P is on d, then it coincides with D and $Pd = 0$.

Consider now the locus of a point P equidistant from two fixed lines f and d. Suppose f and d intersect at a point O. Then the locus consists of the two bisectors of the angles formed by f and d at O. Suppose f and d are parallel. The locus is the line parallel to them and halfway between them. In the projective plane, f and d intersect at their common ideal point, and once again, the ideal line forms the second branch of the locus.

Let F and D be fixed points. What is the locus of a point P such that $\frac{PF}{PD} = e$, where e is an arbitrary positive constant not equal to 1? This constant is referred to as the *eccentricity* of the locus. We may assume that $e < 1$. If $e > 1$, just interchange the labels F and D.

This locus intersects the line DF at two points, V between D and F and V' beyond DF. They can be located as follows. Construct a triangle PDF where $PD = DF$ and $PF = eDF$. Let Z be a point on the extension of DP. Then the bisectors of $\angle DPF$ and $\angle FPZ$ intersects the line DF at V and V' respectively. By the Angle Bisector Theorem, $\frac{VF}{VD} = \frac{PF}{PD} = e = \frac{V'F}{V'D}$.

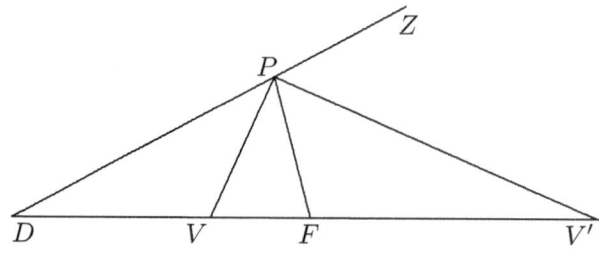

Figure 3.3

Note that V and V' are fixed points determined solely by F, D and e and independent of the choice of P. In fact, they are harmonic conjugates with respect to F and D.

We claim that the desired locus is the circle with diameter VV'. This is known as an *Apollonian* circle.

First, we prove that every point P on the locus indeed lies on this circle. That $\angle VPV' = \angle VPF + \angle FPV' = \frac{1}{2}(\angle DPF + \angle FPZ) = 90°$ follows from $\frac{PF}{PD} = e = \frac{VF}{VD} = \frac{V'F}{V'D}$. Hence P lies on the circle with diameter VV'.

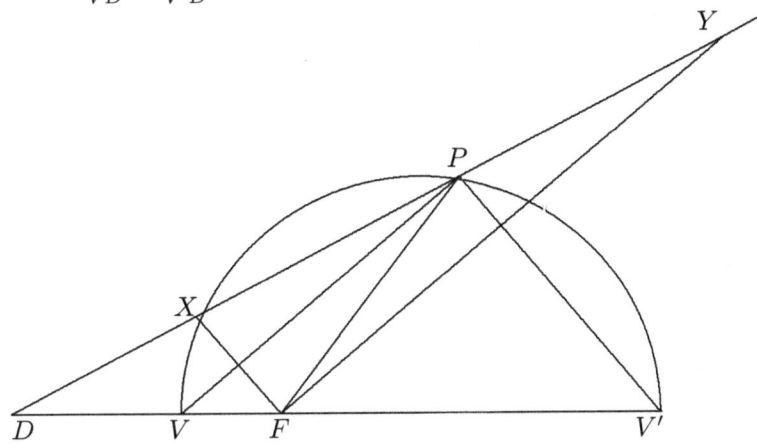

Figure 3.4

Now we prove that every point on this circle lies on the locus. Let P be any point on the circle. From F, draw lines parallel to PV' and PV, cutting the line PD at X and Y, respectively. Since $\angle VPV' = 90°$, $\angle XFY = 90°$ also. By similar triangles, $\dfrac{DP}{PY} = \dfrac{DV}{VF} = e = \dfrac{DV'}{FV'} = \dfrac{DP}{PX}$. Hence $PX = PY$. Being the midpoint of XY, P is the center of the circle with diameter XY, which passes through F. Hence $PF = PX = PY$ so that indeed $\frac{PF}{PD} = e$.

Let f and d be fixed lines. What is the locus of a point P such that $\frac{Pf}{Pd} = e$, where e is an arbitrary positive constant not equal to 1? In the case where f and d are parallel, it consists of two lines parallel to them, one between them and the other beyond them. In the case where f and d intersect, the locus consists of two lines through their point of intersection. The distinction between these two cases vanishes in the projective plane.

Let F be a fixed point and d a fixed line. What is the locus of a point P such that $\frac{PF}{Pd} = e$, where e is an arbitrary positive constant? These loci are not encountered in elementary Euclidean geometry. These are called the *conic sections*, and are studied in projective or analytic geometry. In this chapter, we take an Euclidean look at them.

Consider a hollow infinite double cone whose axis is vertical and whose horizontal cross-sections are circles. The intersection of this surface with a plane is naturally called a conic section. What the curve actually is depends on the relative position of the plane and the double cone.

A line on the double cone must pass through its vertex, and may be considered as a *generator* of the double cone in that if we revolve this line about the vertical axis, the double cone is generated. Let θ be the angle made by a generator of the double cone with the horizontal plane, and let ϕ be the angle made by a plane with the horizontal plane.

Suppose $\phi > \theta$. The plane may pass through the vertex, and the intersection is a single point. Otherwise, it intersects only one part of the double cone. The closed curve obtained as the intersection is called an **ellipse**. A special case is when the plane is itself horizontal, and the ellipse becomes a circle. As mentioned before, a single point is a degenerate case of an ellipse.

When $\phi = \theta$ and the plane passes through the vertex, the intersection is a generator. Otherwise, it intersects only one part of the double cone. The open curve obtained as the intersection is called a **parabola**, with a single line as a degenerate case. Since all parabolas are intersections of the double cone with parallel planes, the non-degenerate ones are all similar to each other.

When $\phi > \theta$ and the plane passes through the vertex, the intersection is a pair of generators. Otherwise, it intersects both parts of the double cone, yielding an open curve in each part. The two branches together form a **hyperbola**, and a degenerate hyperbola is a pair of intersecting straight lines.

In the projective plane, all three loci become single closed curves, because the two open ends of a parabola join up at infinity, and the four open ends of a hyperbola join up in two pairs in a crossing pattern.

In our Euclidean study of the conic sections, they are defined as loci of a point P with respect to a fixed point F and a fixed line d. If $\frac{PF}{Pd} = 1$, we have a parabola. This will be studied in the next section. If $\frac{PF}{Pd} < 1$, we have an ellipse. If $\frac{PF}{Pd} > 1$, we have a hyperbola. These will be studied in the last section.

Exercises:

1. A certain ride in an amusement park consists of two intersecting rods pivoted at two fixed points A and B. They rotate back and forth at the same constant speed in the same direction, like a pair of windshield wipers. The direction is reversed simultaneously for both rods if and only if either presses against the other pivot. A sliding cup C is always at the intersection of the two rods, long enough so that the cup does not slide off their ends. What is the locus of the point C during this motion?

2. Another ride in the amusement part consists of a circular disk rolling along the inside edge of a circular ring twice the diameter of that of the disk. There is no skidding between the disk and the ring. A cup C is at a fixed point on the circumference of the disk. Initially, it coincides with a fixed point A on the circumference of the ring. What is the locus of the point C during this motion?

3. A non-horizontal plane intersects one part of a hollow infinite double cone whose axis is vertical and whose horizontal cross-sections are circles. There are two spheres which are tangent to both the plane and the double cone. The region in the plane inside the double cone is bounded by an ellipse, and this region may be considered as the shadow of the smaller sphere from a light source at the vertex of the double cone. Prove that the sum of the distances from any point on the ellipse to the two points of tangency of the spheres with the plane is constant.

Section 2. The Parabola.

In this section, we move away from the double cone and define a parabola as the locus of a point P equidistant from a fixed point F and a fixed line d. We call F the *focus* and d the *directrix* of the parabola. The line a through F perpendicular to d is called the *axis* of the parabola, and we denote by D the point of intersection of d with a.

In the special case where F coincides with D, the locus is simply the line a. This is the degenerate parabola which we have encountered in the last section. Henceforth, we assume that F and D are distinct points. For convenience of description, we take d to be horizontal and F to be above it. Note that the midpoint V of FD lies on the parabola, and is called its *vertex*.

From this definition of the parabola, it is also clear that all parabolas are similar to one another. Sometimes, appearance can be deceptive, as some parabolas appear to be much narrower than others. However, a picture of a small part of an elephant looks very different from a picture of a small replica of the elephant. See also Exercise 4.

With only the Euclidean tools of the straight-edge and compass, it is impossible to draw the whole parabola in one stroke. What we can accomplish is the following. For any straight line, we can construct the common points of the line and the parabola, if any.

The horizontal line through V is tangent to the parabola at V. A horizontal line below V will be disjoint from the parabola. A horizontal line h above V will intersect the parabola in two points. They can be constructed as in Figure 3.5.

Let N be a point on a above V and h be the horizontal line through N. Then $ND > NF$, so that a circle with center F and radius ND will intersect h at two points P and P'. We have $PF = Pd$ and $P'F = P'd$, so that P and P' are the points of intersection of h with the parabola.

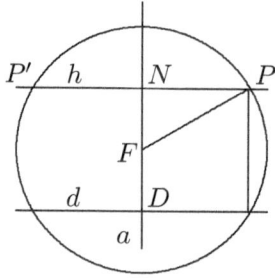

Figure 3.5

Note that P and P' are always symmetric about a. Hence a is an axis of symmetry of the parabola. The distance between P and P' increases with the distance between h and d. Denote by N the point of intersection of h with a. By Pythagoras' Theorem,

$$PN^2 = FP^2 - FN^2 = PM^2 - FN^2 = DN^2 - (DN - DF)^2 = DF(2DN - DF).$$

Since DF is constant, PN will increase with DN.

The situation with vertical lines is even simpler. The axis a intersects the parabola at just the vertex V. Let v be any other vertical line, intersecting d at a point M different from D. The line FM is never horizontal. Hence the perpendicular bisector of the segment FM is never vertical, and intersects v at some point P. We have $PF = PM = Pd$, and P is the only point on the parabola which lies on v. This is because $DMPN$ is a rectangle. Since PN increases with DN, PM increases with DM.

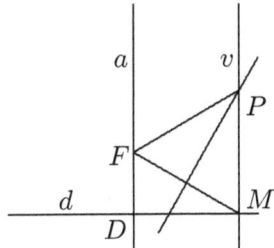

Figure 3.6

We now turn our attention to lines which are neither horizontal nor vertical. To begin with, consider the special case where such a line ℓ passes through the focus F. Let L be the point of intersection of ℓ and the directrix d. Take any point O on ℓ and let the vertical line through O intersect d at T. Construct the circle with center O and radius OT, cutting ℓ at E between L and O and E' on the extension of LO. Then OTE and OTE' are isosceles triangles.

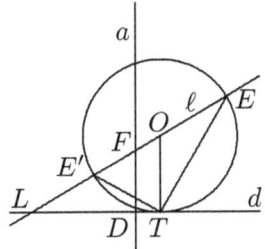

Figure 3.7

From F, draw lines parallel to TE' and TE, intersecting d at M and M' respectively. Draw vertical lines through M and M', intersecting ℓ at P and P' respectively. Then $P'M'F$ and PMF are isosceles triangles similar to OTE and OTE' respectively. Hence $Pd = PF$ and $P'd = P'F$, so that P and P' are the points of intersection of ℓ with the parabola.

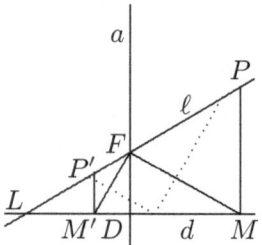

Figure 3.8

The above method generalizes to the case where the line ℓ does not pass through F. Let O be an arbitrary point on ℓ and above d. With O as center, draw a circle tangent to d at T. Now the line FL intersects this circle in zero, one or two points. Figure 3.9 illustrates the case where LF is tangent to the circle, so that there is exactly one such point E. Then triangle TOE is an isosceles triangle.

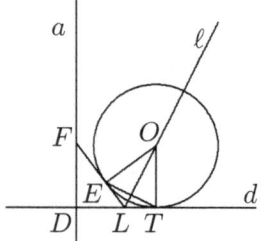

Figure 3.9

From F, draw a line parallel to TE, intersecting d at M. Draw the vertical line through M, intersecting ℓ at P. Then PMF is an isosceles triangle similar to OTE. Hence $Pd = PF$, so that P is the point of intersection of ℓ with the parabola.

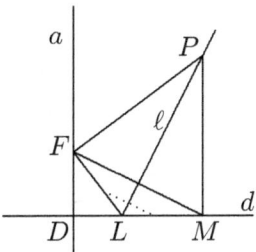

Figure 3.10

Putting together what we have learned from the intersections of various lines with a parabola, we may expect that a parabola looks like the curve in Figure 3.11. It divides the plane into two regions. We shall call the region containing F the *inside* of the parabola, and the other region the *outside*. Points inside the parabola are closer to F than to d, while points outside the parabola are closer to d than to F.

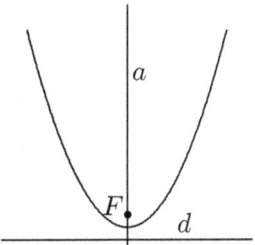

Figure 3.11

In Figure 3.10, ℓ is the perpendicular bisector of FM. For any point Q on ℓ other than P, QM is not vertical. Hence $QF = QM > Qd$, so that Q is outside the parabola. This means that ℓ is a tangent to the parabola.

Every point on the parabola has a unique tangent. The horizontal line through the vertex V is the unique tangent to the parabola at V. For any point P on the parabola other than V, let the vertical line through P intersect d at M. Then the perpendicular bisector of the segment FM is tangent to the parabola at P.

We must still prove that no other line through P can be tangent to the parabola. Let O be the midpoint of PF and G be the midpoint of FM. Then G lies on the circle with diameter PF since $\angle FGP = 90°$. Moreover, G is the lowest point of this circle since OG is a vertical line.

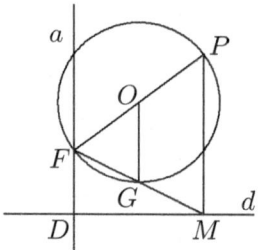

Figure 3.12

Suppose a line ℓ' through P but not G is also a tangent to the parabola. Let the line through F perpendicular to ℓ' intersect d at M'. Let G' be the midpoint of FM'. Since GG' is horizontal, G' is outside the circle with diameter PF. Since ℓ is a tangent, every point on the line g through G' parallel to ℓ' is outside the parabola. However, the point P' of intersection of g with the vertical line through M' lies on the parabola since $P'F = P'M'$. This is a contradiction.

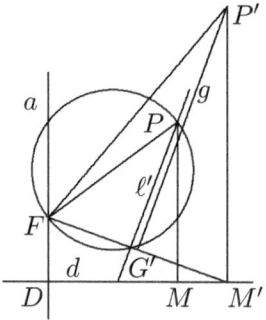

Figure 3.13

The parabola has a significant presence in everyday life. When a light ray reflects off a straight mirror, the angle of incidence is equal to the angle of reflection. If it hits a point on a curved mirror, the tangent to the curve at the point will act as a straight mirror. If we twirl the parabola about its axis into the curved surface called a *paraboloid*, and coat the inside with reflecting material, we have a parabolic mirror.

If a light source is placed at the focus, the light rays will reflect off the parabolic mirror in a direction parallel to the axis. This will give a parallel beam of light, instead of having it scattered off in all directions. Thus the parabolic mirror is very useful in constructing automobile headlights, searchlights and telescopes, among other things. See Exercise 5.

When a projectile is launched under only the influence of gravity, without air or other resistance, its trajectory is a parabola.

Exercises:

4. Figure 3.14 shows two parabolas with a common axis a and a common vertex V. The first one has focus F and directrix d while the second has focus F' and directrix d'. The directrices d and d' intersect a at D and D', respectively. The five points marked by black dots are F', F, V, D and D' from top to bottom. Then $\frac{VF'}{VF} = \frac{VD'}{VD} = r$ for some constant ratio $r > 1$. For any point P on the first parabola, let P' be the point of intersection of the line VP with the second parabola. Prove that $\frac{VP'}{VP} = r$.

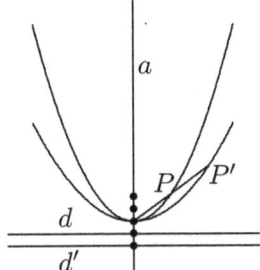

Figure 3.14

5. A light source is placed at the focus of a parabola. Prove that a light beam reflected off the parabola is parallel to the axis of the parabola.

6. From a point T outside a parabola with focus F, tangents are drawn to the parabola at the points P and P'. Prove that $\angle TFP = \angle TFP'$.

Section 3. Ellipses and Hyperbolas.

Let F be a fixed point F and d be a fixed line d. In the preceding section, we define a parabola as the locus of a point P such that $\frac{PF}{Pd} = 1$. In this section, we replace the constant 1 by some positive number $e \neq 1$. Because of the non-symmetry between F and d, the nature of the locus depends on whether $e < 1$ or $e > 1$. The results are the other two types of conic sections. If $e < 1$, the curve is called an *ellipse*. If $e > 1$, the curve is called a *hyperbola*.

The point F is called a *focus*, the line d a *directrix* and the constant e the *eccentricity* of the curve. The line a through F perpendicular to d is called the *major axis* of the curve, and we denote by D the point of intersection of d with a.

Consider the special case where F coincides with D. If $e < 1$, the locus consists only of the point D. When $e > 1$, the locus consists of a pair of lines through D, making equal angles with d. These are the degenerate forms of an ellipse and a hyperbola, respectively. Henceforth, we assume that F and D are distinct points.

With only the Euclidean tools of the straight-edge and compass, it is impossible to draw either an ellipse or a hyperbola, except in the special case where the ellipse is actually a circle. As in the last section, for any straight line, we shall construct the common points of the line and the curve, if any.

Our discussion will be focused on ellipses. For convenience, we take d to be horizontal and F to be above it.

In each of the constructions below, the points of intersection of the ellipse with a specific line must also be the points of intersection of that line with some circle. Since a line and a circle intersect in 0, 1 or 2 points, we have indeed found all of them. We will not repeat this for each construction.

We first consider the points of intersection of our ellipse with vertical lines. On the axis a, construct the Apollonius circle with respect to D, F and e, intersecting a at V and V', with V closer to d. Since $\frac{VF}{Vd} = \frac{VF}{VD} = e$ and $\frac{V'F}{V'd} = \frac{V'F}{V'D} = e$, these points lie on the ellipse, and are called its *vertices*.

Now let v be any vertical line ℓ other than a, intersecting d at the point M. Construct the Apollonius circle with respect to F, M and e. It intersects v in 0, 1 or 2 points. Suppose P is such a point. Then $\frac{PF}{Pd} = \frac{PF}{PM} = e$, and P lies on our ellipse.

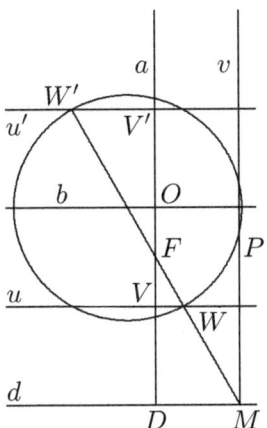

Figure 3.15

Let this Apollonius circle intersect FM at W and W', with W closer to M. Since $\frac{WF}{WM} = e$, triangles FVW and FDM are similar, so that W lies on the horizontal line u through V. Similarly, W' lies on the horizontal line u' through V'. Since the center of this circle is the midpoint of WW', it lies on the horizontal line b through the midpoint O of VV'. It follows that the ellipse is symmetric about b. We call b the *minor axis* of the ellipse.

We next consider the points of intersection of our ellipse with any horizontal line h. We draw a circle with center F and radius equal to e times the distance between d and h. It intersects h in 0, 1 or 2 points. Suppose P is such a point. Then $\frac{PF}{Pd} = e$ and P lies on our ellipse. From this, it is immediately clear that our ellipse is symmetric about a. Thus we have established that the ellipse is a centrally symmetric figure. We call O the *center* of the ellipse.

Note that V lies on a. Hence it is the only point on u that belongs to the locus. Similarly, V' is the only point on u' that belongs to the locus. Thus we expect that the entire ellipse lies between these two lines. Indeed, for any point P on any horizontal line below u, we have $\frac{PF}{Pd} > \frac{VF}{Vd} = e$. Now let P be a point on any horizontal line above u'. Let PF intersect u' at Q. Then $\frac{PF}{Pd} = \frac{QF}{Qd} > \frac{V'F}{V'd} = e$. Thus the vertical dimension of the ellipse is finite. It follows immediately that so is its horizontal dimension.

Finally, we consider the points of intersection of our ellipse with any line which is neither vertical nor horizontal. Let ℓ be such a line passing through F, intersecting d at a point L other than D. Construct the Apollonius circle with respect to F, L and e and let it intersect ℓ at P and P'. These are the points of intersection of ℓ with the ellipse.

Suppose ℓ does not pass through F. Let it intersect d at a point L. Let Y be the point on ℓ at a distance 1 above d. Draw a circle with center Y and radius e. It may intersect LF in 0, 1 or 2 points. Suppose X is such a point. Let the line through F parallel to XY intersect ℓ at P. Then $\frac{PF}{Pd} = \frac{YX}{Yd} = e$, and P lies on the ellipse.

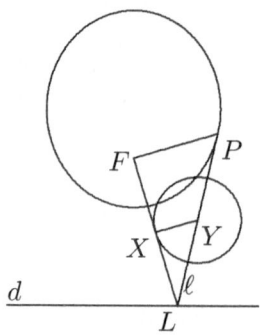

Figure 3.16

We now establish two important properties of tangents to an ellipse, analogous to those for the parabola in Exercise 5. In the preceding construction, suppose X is the only point of intersection of LF with the circle centered at Y. Then LF is a tangent to this circle, and is perpendicular to the radius XY. Now ℓ is a tangent to the ellipse at P. Since PF is parallel to XY, $\angle PFL = 90°$. In other words, *the part of every tangent between the point on the ellipse and the point on the directrix subtends a right angle at the focus.*

Let F' and D' be the points symmetric to F and D respectively, with respect to O, and let d' be the horizontal line through D'. Then the same ellipse is defined if we use F' as the focus and d' as the directrix. Hence an ellipse has two foci and two directrices. In the projective plane, the other focus of a parabola is the ideal point of its axis, and the other directrix is the ideal line.

Suppose ℓ is tangent to the ellipse at P and intersects d and d' respectively at L and L'. Let the vertical line through P intersect d and d' respectively at M and M'. Then triangles PML and $PM'L'$ are similar, so that $\frac{PL}{PL'} = \frac{PM}{PM'} = \frac{PF}{PF'}$. Since $\angle PFL = 90° = \angle PF'L'$, triangles PFL and $PF'L'$ are also similar. This follows from the rarely used *RHS* case for similarity. Hence $\angle FPL = \angle F'PL'$. In other words, *every tangent of an ellipse makes equal angles with the lines joining the point of tangency to the two foci.*

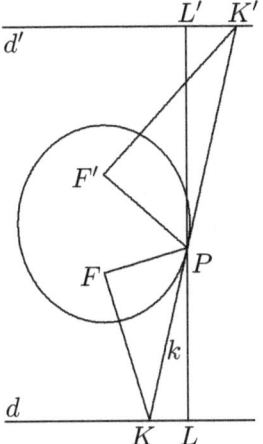

Figure 3.17

If a light source is at one focus of an elliptical mirror, all the reflected rays converge on the other focus.

Let P be an arbitrary point on the ellipse. Let the vertical line through P intersect d and d' at M and M' respectively. Then $\frac{PF}{PM} = e = \frac{PF'}{PM'}$. Hence $PF + PF' = e(PM + PM') = eMM'$, which is a constant. This is an alternative definition of the ellipse, as the locus of a point whose distances from two fixed foci have constant sum.

With this alternative definition of an ellipse, we see easily how a circle is a special case of an ellipse. The two foci coincide and becomes the center of the circle while both directrices recede to coincide as the ideal line in the projective plane.

Using the new definition, we can still prove that *every tangent of an ellipse makes equal angles with the lines joining the point of tangency to the two foci.*

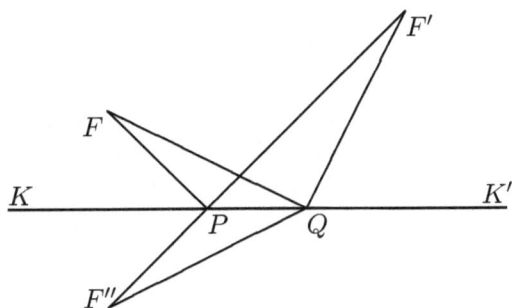

Figure 3.18

Let F and F' be the foci and LL' be a tangent to the ellipse. Let F'' be the reflection of F across LL'. We claim that the point P of intersection of LL' with $F'F''$ is the point of tangency.

Since exactly one point on LL' is on the ellipse while every other point on LL' is outside the ellipse, we only need to prove that $PF+PF' < QF+QF'$ for any other point Q on LL'. It follows from the Triangle Inequality that $QF + QF' = QF'' + QF' > F''F' = PF'' + PF' = PF + PF'$.

The new definition of an ellipse does not mention the directrices at all. Thus it is impossible for us to prove from it that *the part of every tangent between the point on the ellipse and the point on the directrix subtends a right angle at the focus.* Instead, we shall make use of this to define the directrices, and show that the old definition also follows from the new one.

Let P be any point on the ellipse and let ℓ be the tangent to the ellipse at P, Let L and L' be points on it such that $\angle PFL = 90° = \angle PF'L'$. Let d and d' be the horizontal lines through L and L', cutting the vertical line through P at M and M' respectively. Let the lines through F and F' perpendicular to ℓ cut LL' at Q and Q', and MM' at R and R', respectively.

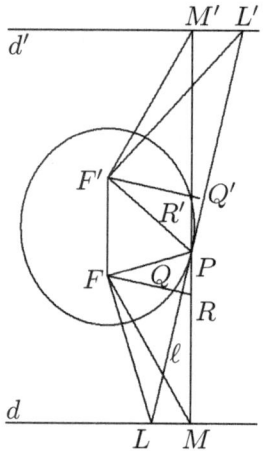

Figure 3.19

We claim that d and d' are the desired directrices. For justification, we need to prove first that their choices do not depend on the specific point P on the ellipse. Then we have to show that $\frac{PF}{Pd}$ is constant for all pints P on the ellipse.

Since $\angle FPL = \angle F'PL'$ and $\angle PFL = \angle PF'L' = \angle PQF = \angle PQ'F' = 90°$, triangles PFL, $PF'L'$, PQF and $PQ'F'$ are all similar to one another. Hence $\frac{PK}{PF} = \frac{PK'}{PF'} = \frac{PF}{PQ} = \frac{PF'}{PQ'}$. Denote the value of this common ratio by μ. Then $\frac{LL'}{PF+PF'} = \frac{PL+PL'}{PF+PF'} = \mu$ and $\frac{QQ'}{PF+PF'} = \frac{PQ+PQ'}{PF+PF'} = \frac{1}{\mu}$. Multiplication yields $LL' \cdot QQ' = (PF + PF')^2$.

Since $\angle MPL = \angle M'PL'$ and $\angle PML = \angle PM'L' = \angle PQR = \angle PQ'R' = 90°$, triangles PML, $PM'L'$, PQR and $PQ'R'$ are all similar to one another. Hence $\frac{PL}{PM} = \frac{PL'}{PM'} = \frac{PR}{PQ} = \frac{PR'}{PQ'}$. Denote the value of this common ratio by ν. Then $\frac{LL'}{MM'} = \frac{PL+PL'}{PM+PM'} = \nu$ and $\frac{QQ'}{RR'} = \frac{PQ+PQ'}{PR+PR'} = \frac{1}{\nu}$. Multiplication yields $LL' \cdot QQ' = MM' \cdot RR'$.

It follows that $(PF + PF')^2 = MM' \cdot RR'$. Since $FF'R'R$ is a parallelogram, $RR' = FF'$. Hence $MM' = \frac{(PF+PF')^2}{FF'}$ has constant length. Since $\angle PFL = \angle PML = \angle PF'L' = \angle PM'L' = 90°$, $FLMP$ and $F'L'M'P$ are cyclic quadrilaterals. Hence $\angle FMP = \angle FLP = \angle F'L'P = \angle F'M'P$, so that $FF'M'M$ is an isosceles trapezoid. Thus d and d' are situated symmetrically about the ellipse, and since the distance between them is constant, their locations are uniquely determined, independent of the choice of the point P on the ellipse.

From the similar triangles, we have $\frac{PF}{PF'} = \frac{PL}{PL'} = \frac{PM}{PM'}$. It follows that $\frac{PF}{PM} = \frac{PF'}{PM'} = \frac{PF+PF'}{MM'}$ is constant, and the ellipse does satisfies the original definition as well.

We now return to the problem of the cat on the sliding ladder. For definiteness, take C to be the point on AB such that $AC = 2BC$. Let another cat sit on the midpoint C' of a longer ladder $A'B'$ where we have $A'C' = B'C' = AC = 2BC$. Initially, the height of C above the ground is half that of C'. Let the ladders slide so that they are always parallel to each other. Then $A'C'CA$ is a parallelogram so that CC' is vertical. Let it intersect the floor at a point M. Then triangles MBC and $MB'C'$ are similar, so that $MC' = 2MC$.

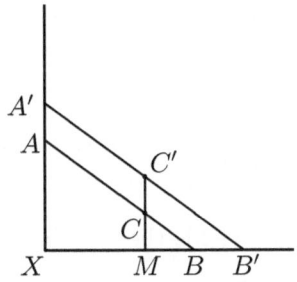

Figure 3.20

Since the locus of C' is a quarter of a circle, the locus of C is a quarter of a squashed circle, where each point is half its original distance from a fixed diameter of the circle. A squashed circle looks like an ellipse, and we now prove that it is indeed an ellipse. Unfortunately, we have to resort to heavy trigonometric computations. An Euclidean argument is much desired.

Let F and F' be fixed points and α a fixed length. By our new definition, the locus of a point P such that $PF + PF' = 2\alpha$ is an ellipse with foci F and F'. The points on this ellipse can be generated as follows. Take an arbitrary point S satisfying $SF' = 2a$ and let T be the midpoint of SF. Then the point P of intersection of SF' and the line through T perpendicular to SF lies on the ellipse since $PF + PF' = PF' + PS = SF' = 2\alpha$.

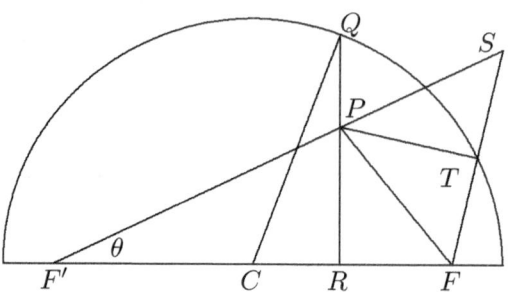

Figure 3.21

Let $FF' = 2\sqrt{\alpha^2 - \beta^2}$ for a fixed length $\beta < \alpha$ and let O be the midpoint of FF'. Let the line through P perpendicular to FF' cut FF' at R, and the circle with center O and radius α at Q. We claim that $\frac{QR}{PR} = \frac{\alpha}{\beta}$. In other words, the ellipse may be obtained from the circle by contracting the points on the circle towards a diameter by a fixed ratio.

To prove this equality, we establish a coordinate system with O as the origin and the ray OF as the positive x-axis. Then the coordinates of F and F' are $(\pm\sqrt{\alpha^2 - \beta^2}, 0)$ respectively. Let θ denote $\angle SF'F$. Then the coordinates of S are $(2\alpha\cos\theta - \sqrt{\alpha^2 - \beta^2}, 2\alpha\sin\theta)$. Hence the coordinates of T are $(\alpha\cos\theta, \alpha\sin\theta)$. The equations of SF' and PT are respectively

$$\frac{y}{x + \sqrt{\alpha^2 - \beta^2}} = \frac{\sin\theta}{\cos\theta} \quad \text{and} \quad \frac{y - \alpha\sin\theta}{x - \alpha\cos\theta} = \frac{\sqrt{\alpha^2 - \beta^2} - a\cos\theta}{\alpha\sin\theta}.$$

Solving this system of equation yields the coordinates of P, which are

$$x = \frac{a^2\cos\theta - a\sqrt{a^2 - b^2}}{a - \sqrt{a^2 - b^2}\cos\theta} \quad \text{and} \quad y = \frac{b^2\sin\theta}{a - \sqrt{a^2 - b^2}\cos\theta}.$$

$$\text{Now } QR = \sqrt{QC^2 - RC^2}$$

$$= \sqrt{a^2 - \left(\frac{a^2\cos\theta - a\sqrt{a^2-b^2}}{a - \sqrt{a^2-b^2}\cos\theta}\right)^2}$$

$$= \frac{a}{a - \sqrt{a^2-b^2}\cos\theta}\sqrt{(a^2 - (a^2-b^2))(1 - \cos^2\theta)}$$

$$= \frac{ab\sin\theta}{a - \sqrt{a^2-b^2}\cos\theta}.$$

Thus we indeed have $\frac{QR}{PR} = \frac{a}{b}$, so that an ellipse is a squashed circle. We now establish that a squashed circle is an ellipse.

Let O be the center of a circle with radius α. Let U be a point such that $CU = \beta$. Let F and F' be points on the diameter of the circle perpendicular to CU, such that $UF = UF' = \alpha$. Let Q be an arbitrary point on the circle. Let R be the foot of perpendicular from Q to FF', and let P be the point on QR such that $\frac{QR}{PR} = \frac{\alpha}{\beta}$. We claim that $PF + PF' = 2\alpha$.

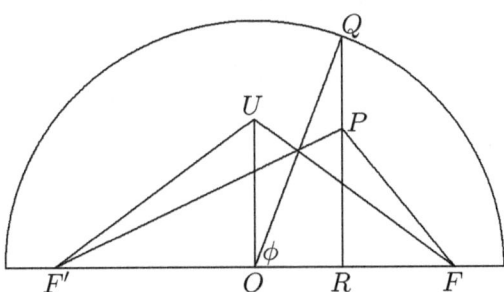

Figure 3.22

Let ϕ denote $\angle QOR$. Then the coordinates of Q are $(\alpha\cos\phi, \alpha\sin\phi)$. Hence the coordinates of P are $(\alpha\cos\phi, \beta\sin\phi)$. Now

$$(PF+PF')^2 = \left(\sqrt{\alpha\cos\phi + \sqrt{\alpha^2 - \beta^2})^2 + r^2\sin^2\phi}\right.$$

$$\left. + \sqrt{(\alpha\cos\phi - \sqrt{\alpha^2 - \beta^2})^2 + r^2\sin^2\phi}\right)^2$$

$$= 2\alpha^2\cos^2\phi + 2(\alpha^2 - \beta^2) + 2\beta^2\sin^2\phi$$

$$+ 2\sqrt{(\alpha^2\cos^2\phi + \alpha^2 - \beta^2 + \beta^2\sin^2\phi)^2 - 4\alpha^2(\alpha^2 - \beta^2)\cos^2\phi}$$

$$= 2\alpha^2\cos^2\phi + 2(\alpha^2 - \beta^2) + 2\beta^2\sin^2\phi + 2\alpha^2\sin^2\phi + 2\beta^2\cos^2\phi$$

$$= 4\alpha^2.$$

A treatment of the hyperbola is analogous to be the above treatment of an ellipse. In the exercises, we investigate some properties of a hyperbola. Figure 3.23 shows in dots a typical hyperbola. It may be defined by the focus-directrix pair (F, d) or (F', d'). Like an ellipse, it has bilateral symmetry about each of two perpendicular axes. The two slanting lines in Figure 3.23 are called *asymptotes*, and may be thought of as tangents at their respective ideal points.

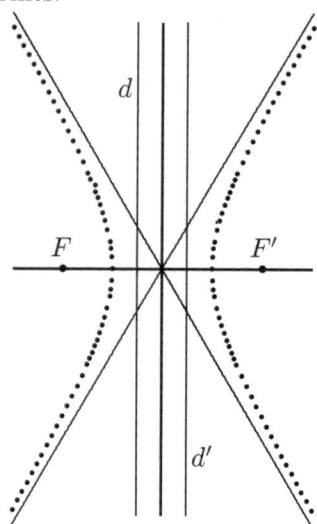

Figure 3.23

Exercises:

7. A plane intersects both parts of a hollow infinite double cone whose axis is vertical and whose horizontal cross-sections are circles. There are two spheres which are tangent to both the plane and the double cone, on the same side of the plane as the vertex of the double cone. The shadow of the spheres on the plane from a light source at the vertex of the double cone consists of two parts, each bounded by a branch of a hyperbola. Prove that the difference of the distances from any point on the hyperbola to the two points of tangency of the spheres with the plane is constant.

8. Prove that if a segment joins a point of a hyperbola to a point on the directrix, the segment subtends a right angle at the corresponding focus.

9. A light source is placed at one of the foci of a hyperbola. Prove that a light beam reflected off the hyperbola appears to be emitted from the other focus.

Bibliography

[1] R. Barrington Leigh, I. E. Leonard, J. E. Lewis, Andy Liu and G. Tokarsky, Ellipses, *Mathematical Wizardry for a Gardner*, edited by Ed Pegg Jr., Alan H. Schoen and Tom Rodgers, A K Peters, Wellesley (2009) 221–234.

[2] Martin Gardner, *New Mathematical Diversions*, Mathematical Association of America, Washington (1995) 173–183.

[3] Martin Gardner, *Penrose Tiles to Trapdoor Ciphers*, Mathematical Association of America, Washington, (1997) 205–218.

[4] Martin Gardner, *Last Recreations*, Springer, New York, (1997), 285–301.

[5] V. L. Gutenmacher and N. B. Vasilyev, *Straight Lines and Curves*, Mir Press, Moscow (1980) 1–8.

[6] I. E. Leonard, J. E. Lewis, Andy Liu and G. Tokarsky, The Parabola — An Euclidean Introduction for Smart Novices, Mathematics and Informatics Quarterly, **6** (1996) 122–131.

Solutions to Exercises

1. Note that $\angle CAB = \theta$ is constant. Hence the locus of C is the part of the circle from A to B subtending an angle of θ.

2. Note that as long as the disk is in contact with the ring, the center O of the ring will lie on the circumference of the disk. We may assume that during the rotation, the disk moves counterclockwise. Figure 3.24 shows a general position during the motion when P is the point of tangency of the disk with the ring. The center Q of the disk is the midpoint of OP. Let $\angle COQ = \theta$. Then $\angle OCQ = \theta$ and $\angle CQP = 2\theta$. Hence the length of the arc CP is $QP \cdot 2\theta = OP \cdot \theta$. Since there is no skidding, the arc AP has the same length. This means that $\angle AOP = \theta$, so that C lies on OA. It follows that the locus of C is the diameter of the ring passing through A.

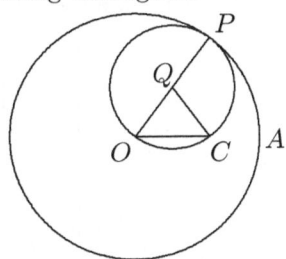

Figure 3.24

3. Every generator of the double cone is tangent to both spheres, and the distance between the two points of tangency is constant regardless of the choice of the generator. Consider the generator passing through a point P on the ellipse. Let Q and Q' be the points of tangency of this generator with the smaller sphere and the larger sphere respectively. Let F and F' be the points of tangency of the plane with the smaller sphere and the larger sphere respectively. It follows that $PQ = PF$ and $PQ' = PF'$, so that $PF + PF' = PQ + PQ' = QQ'$ is constant.

4. Let us get P' in a different way. Draw a vertical line through P, cutting d at M. Extend VM to cut d' at M'. Draw a vertical line through M', cutting the extension of VP at P'. Now triangles VDM and $VD'M'$ are similar, as are triangles VMP and $VM'P'$. Thus $\frac{VF'}{VF} = \frac{VD'}{VD} = \frac{VM'}{VM} = \frac{VP'}{VP}$. It follows that triangles VFP and $VF'P'$ are also similar, so that $\frac{PF'}{P'F'} = \frac{VP'}{VP} = \frac{P'M'}{PM}$. Since P is on the first parabola, $PF = PM$. Hence $P'F' = P'M'$, so that P' indeed lies on the second parabola.

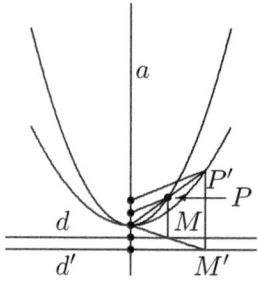

Figure 3.25

5. Let a light beam from the focus F of a parabola reflect off a point P on the parabola. Let the vertical line through P intersect the directrix d at the point M. Then the perpendicular bisector TG of FM is the tangent to the parabola at P. It is also the bisector of $\angle FPM$. Hence $\angle RPT = \angle MPG = \angle FPG$, where R is any point on the extension of MP. It follows that the light beam will reflect off P in the direction PR, which is parallel to the axis a of the parabola.

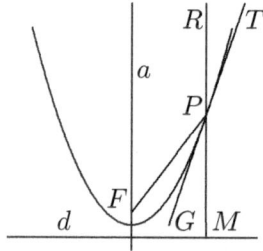

Figure 3.26

6. Let the directrix d intersect PT at L, the vertical line through P at M and the vertical line through T at K. Let W and W' be points on FP and FP' such that TW and TW' are perpendicular to FP and FP', respectively. Since PL is tangent to the parabola, triangles PLM and PLF are congruent. Hence LF is perpendicular to PF and therefore parallel to TW. It follows that $\frac{FW}{FP} = \frac{ZT}{ZP} = \frac{TK}{PM}$. Since $FP = PM$, we have $FW = TK$. Similarly, $FW' = TK$. Hence triangles TWF and triangle $TW'F$ are congruent, so that $\angle TFP = \angle TFP'$.

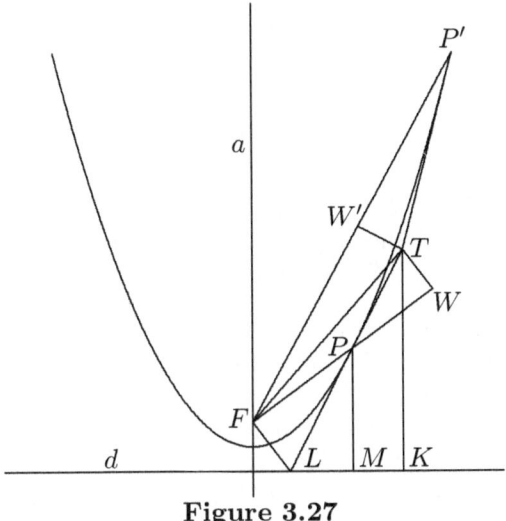

Figure 3.27

7. Every generator of the double cone is tangent to both spheres, and the distance between the two points of tangency is constant regardless of the choice of the generator. Consider the generator passing through a point P on the ellipse. Let Q be the point of tangency of this generator with the sphere in the same part as P and Q' be the point of tangency of this generator with the other sphere. Let F and F' be the points of tangency of the plane with these spheres respectively. It follows that $PQ = PF$ and $PQ' = PF'$, so that $PF' - PF' = PQ' - PQ' = QQ'$ is constant.

8. Let F be the focus of the branch of the hyperbola on which a point P lies. Let the line through F perpendicular to PF intersect the corresponding directrix d at the point L. Let the line through P perpendicular to d intersect it at M. Let T be any point on the line FL other than P. Let the lines through T perpendicular to d and to PF intersect them at K and Q respectively. Since TQ is parallel to FL and TK is parallel to PM, similar triangles yield $\frac{FQ}{FP} = \frac{LT}{LP} = \frac{TK}{PM}$. Now $PF = ePM$ where e is the eccentricity of the hyperbola. Hence $FQ = eTK$.

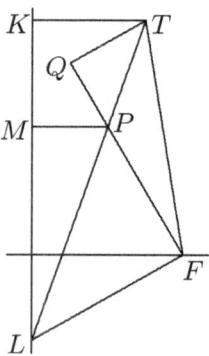

Figure 3.28

Note that $TF > FQ = eTK$ since TF is the hypotenuse of the right triangle FQT. Hence T is not in the region of the hyperbola containing F. It follows that PL is a tangent.

9. Let P be a point on the hyperbola and let the tangent at P intersect the directrices d and d' at L and L' respectively. Let the line through P perpendicular to the directrices intersect them at M and M' respectively. By similar triangles, $\frac{PL}{PL'} = \frac{PM}{PM'}$. Since $PF = ePM$ and $PF' = ePM'$, $\frac{PF}{PF'} = \frac{PL}{PL'}$. By Exercise 8, $\angle PFL = 90° = \angle PF'L'$. Hence triangles PFL and $PF'L'$ are similar, so that $\angle FPL = \angle F'PL'$. When the light beam from F reflects off P, $\angle FPL$ is the angle of incidence and $\angle F'PL'$ is equal to the angle of reflection. Hence the reflected light beam appears to be emitted from F'.

Chapter Four: Inversive Geometry

Section 1. Inversion.

On the Euclidean plane, place a sphere so that its south pole O is at the origin. Let J be the north pole. For any point $Q \neq J$ on the sphere, the point P of intersection of the extension of OQ with the plane is called its image under the **stereographic projection** from J.

It would be nice is there is a point I on the plane which is the image of J. If we add such a point to the Euclidean plane, we will have what is known as the **inversive plane**. Think of the sphere as a balloon and the point J as a puncture. If we stretch the balloon out onto the plane, we can see that the point J is in every direction!

The point I is called the **ideal point** and lies on every straight line. To see this, consider a straight line ℓ on the inversive plane and the plane passing through O and ℓ. The cross-section with the sphere is a circle passing through the point J. This justifies the statement that I lies on every straight line. In fact, it closes the straight line into something like a circle.

Inversion is a transformation of the inversive plane into itself. A point O is chosen as the **center of inversion**. A circle with center O is chosen as the **circle of inversion**. We take O as the south pole of the sphere, and the image of the equator as the circle of inversion. The diameter of the sphere is called the **radius of inversion**, because it is the radius of the circle of inversion.

The diagram below represents a vertical cross-section of the inversive plane and the sphere. The midpoint N of the semicircle OJ is on the equator and its image M is on the circle of inversion. For any point P inside the circle of inversion, let Q be the point on the sphere such that P is the image of Q. Q is necessarily in the southern hemisphere. Let Q' be the image of Q about the plane of the equator, so that it is in the northern hemisphere. Let P' be the image of Q' under stereographic projection. Then P' is necessarily outside the circle of inversion, and is the image of P under inversion.

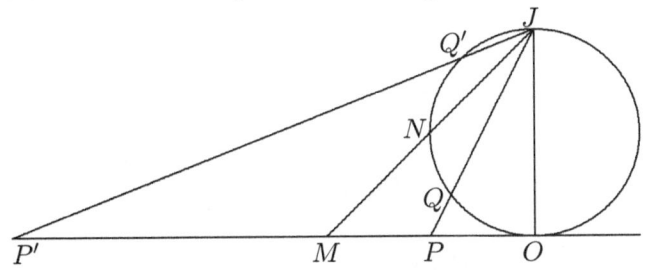

Figure 4.1

© Springer International Publishing AG 2018
A. Liu, *S.M.A.R.T. Circle Minicourses*, Springer Texts in Education, https://doi.org/10.1007/978-3-319-71743-2_4

Note that N is also the midpoint of the arc QQ', so that we have $\angle NJQ = \angle NJQ'$. Hence $\angle OJP = 45° - \angle NJQ = 45° - \angle NJQ' = \angle OP'J$. Now triangles OJP and $OP'J$ are similar, so that $OP \cdot OP = OJ^2 = OM^2$.

Thus it is possible to define inversion without reference to the sphere. Let O be the center of inversion and let r be the radius of inversion. The image of a point P is the point P' on the ray OP such that $OP \cdot OP' = r^2$.

Suppose two line ℓ_1 and ℓ_2 intersect at a point O. Then the angle between them is the directed angle through which ℓ_1 may be rotated about O into ℓ_2. The angle between two parallel lines is taken to be $0°$.

Suppose a line ℓ intersect a circle ω at a point O. Then the angle between them at O is the directed angle through which ℓ may be rotated about O into the tangent t of ω at O. The angle between a line and a circle which are tangent to each other is taken to be $0°$.

Suppose two circles ω_1 and ω_2 intersect at two points O_1 and O_2. Then the angle between them at O_1 is the directed angle through which the tangent t_1 to ω_1 at O_1 may be rotated into the tangent t_2 of ω_2 at O_1 about O_1. The angle between the circles at O_2 will have the same magnitude but opposite orientation. The angle between two tangent circles is taken to be $0°$.

A **conformal** transformation is one in which the magnitudes and orientations of angles are preserved. An **anti-conformal** transformation is one in which the magnitudes of angles are preserved but their orientations are reversed. Like reflection, inversion is an anti-conformal transformation.

To prove this, we first state and prove a simple and yet important result which we will call the **Basic Lemma**. Let P and Q be two points not collinear with the center of inversion O. Let P' and Q' be their respective images. Then triangles OPQ and $OQ'P'$ are similar. The proof is straightforward. By the definition of inversion, we have $OP \cdot OP' = OQ \cdot OQ'$. Hence $\frac{OP}{OQ} = \frac{OQ'}{OP'}$, and the Basic Lemma follows.

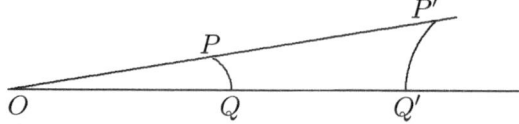

Figure 4.2

Consider now a curve PQ which has been inverted into its image $P'Q'$. The angles between these curves and the radial line OP are formed by the tangents to the curve at P and P' with OP. As the radial line OQ approaches OP, the angle between PQ and OP approaches the angle vertically opposite to $\angle OQP$, while that between $P'Q'$ and OP approaches $\angle OP'Q'$.

By the Basic Lemma, the limiting values are equal. Since triangle OPQ is similar to triangle $OQ'P'$ rather than triangle $OP'Q'$, orientation is reversed. For angles between two arbitrary intersecting curves, just consider the angles they make with the radial line through their point of intersection.

Two curves are said to be **orthogonal** to each other if the angle between them is 90°. Thus orthogonal lines are simply perpendicular lines. A line and a circle are orthogonal if and only if the line passes through the center of the circle. Two circles are orthogonal if the tangent at the point of intersection of each circle passes through the center of the other circle.

Since everything outside the circle of inversion is inverted inside the circle and everything inside the circle of inversion is inverted outside the circle, distances cannot possibly be preserved under inversion. The Basic Lemma provides information on how distances are affected.

From similar triangles, we have

$$\frac{P'Q'}{PQ} = \frac{OQ'}{OP} = \frac{OQ' \cdot OQ}{OP \cdot OQ} = \frac{r^2}{OP \cdot OQ},$$

where r is the radius of inversion. It is easy to prove that the same result holds when O, P and Q are collinear.

We now consider the inversive images of lines and circles with respect to a point O. There are four cases.

Case One.
A line through O inverts into the same line through O.

This follows from the definition of inversion.

Case Two.
A line not through O inverts into a circle through O.

Let Q be the point on the line ℓ such that OQ is perpendicular to the ℓ and let Q' be the inversive image of Q. For any point P on ℓ, let P' be its inversive image. Then triangles OPQ and $OQ'P'$ are similar by the Basic Lemma. Hence $\angle OP'Q' = \angle OQP = 90°$. It follows that P' lies on the circle with diameter OQ'. Conversely, let P' be any point on the circle with diameter OQ', and let P be its inversive image. As before, $\angle OQP = \angle OP'Q' = 90°$, so that P lies on ℓ.

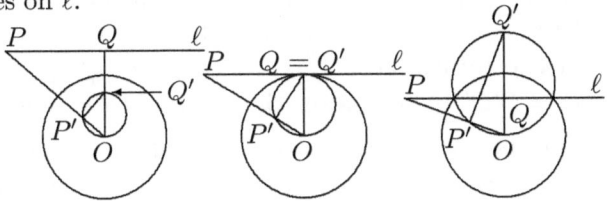

Figure 4.3

Case Three.

A circle through O inverts into a line not through O.

This is proved in exactly the same way as in Case Two.

Case Four.

A circle not through O inverts into another circle not through O.

The proof is similar to that in Case Two.

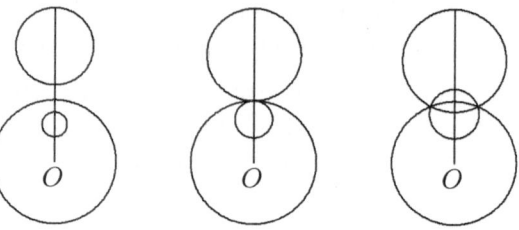

Figure 4.4

What are the inversive images of a square with respect to its incircle and its circumcircle? Since each side does not pass through the center of the circle, they turn into halves of circles passing through the center, inside the incircle and outside the circumcircle.

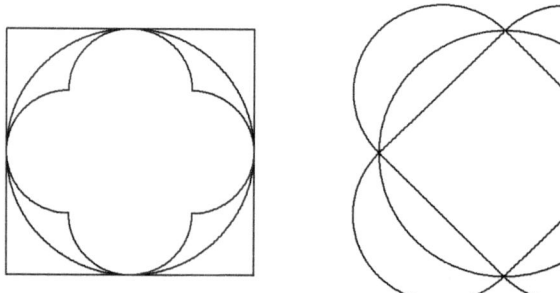

Figure 4.5

Two circles ω and ω' invert into each other with respect to a point O not on either circle. Let PQ be the diameter of ω collinear with O. The midpoint C of PQ is the center of ω. Let the images under inversion be P', Q' and C' respectively. Then $\frac{C'P'}{CP} = \frac{r^2}{OC \cdot OP} \neq \frac{r^2}{OC \cdot OQ} = \frac{C'Q'}{CQ}$. Since $CP = CQ$, $C'P' \neq C'Q'$, which means that C' is not the center of ω'.

Suppose O coincides with P. Then the image of ω is the line ℓ through Q' perpendicular to PQ, which may be considered as a circle with infinite radius centered at the ideal point. However, $C'Q' = \frac{CQ \cdot r^2}{PC \cdot PQ} = \frac{CQ \cdot PQ \cdot PQ'}{PC \cdot PQ} = PQ'$. It follows that C' is the reflection of P across ℓ, and not the ideal point.

What happens if the image w' of the circle w is w itself? Such circles w do exist. There is the circle of inversion itself. However, the inversive image of its center O is not O itself but the ideal point of the inversive plane.

All other circles which invert into themselves are circles orthogonal to the circle of inversion. The image C' of the center C is still not the center of w'. Obviously, C must be outside the circle of inversion, which means that C' must be inside. Hence they cannot coincide.

Exercises:

1. (a) What is the inversive image of the steering wheel in Figure 4.6 on the left with respect to the circle?

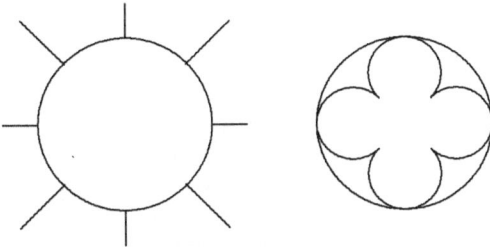

Figure 4.6

   (b) What is the inversive image of a clover leaf in Figure 4.6 on the right with respect to the circle?

2. Prove that the image of a circle orthogonal to the circle of inversion is itself.

3. Let A, B, C and D be four concyclic points. A variable circle through A and B touches a circle through C and D at T. Prove that the locus of T is a straight line or a circle.

Section 2. Applications to Euclidean Geometry.

Problems in Euclidean geometry not involving circles are in general simpler than problems involving circles. Thus the inversive method is very useful in solving Euclidean problems involving circles, by inverting them into straight lines. We illustrate with the following problem.

Three circles ω_1, ω_2 and ω_3 pass through O. A is the other point of intersection of ω_1 and ω_2, B is the other point of intersection of ω_2 and ω_3, and C is the other point of intersection of ω_3 and ω_1. The extension of AO intersects ω_1 again at D, the extension of BO intersects ω_2 again at E, and the extension of CO intersects ω_3 again at F. Prove that if OE and OF are diameters of ω_2 and ω_3 respectively, then OD is a diameter of ω_1.

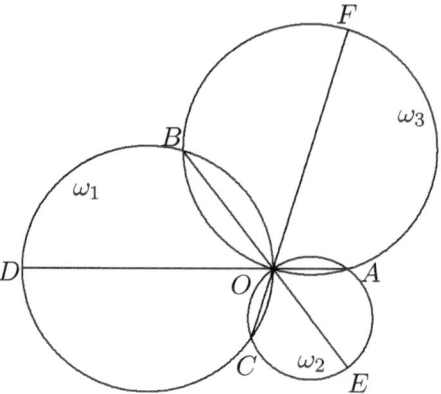

Figure 4.7

To use inversion, we have to choose a center O of inversion. Remember that a circle turns into a straight line if and only if O lies on the circle. Remember that inversion can also change straight lines into circle. This will not happen if O lies on the straight line. So we should choose O as the point through which most of the lines and circles pass.

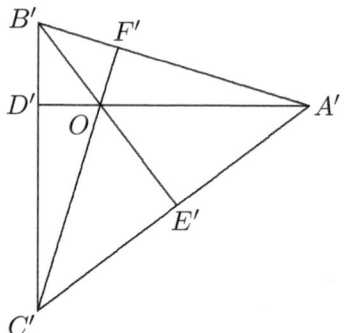

Figure 4.8

In the above problem, the point O itself is the obvious choice. Then the three circles turn into triangle $A'B'C'$ while the radial lines OA, OB and OC turn into themselves. That OE is a diameter of ω_2 means that $B'E'$ is orthogonal to $A'C'$. Similarly, $C'F'$ is orthogonal to $A'B'$. Hence O is the orthocenter of triangle $A'B'C'$, so that $A'O$ is orthogonal to $B'C'$. It follows that OD is indeed a diameter of ω_1.

Another useful point to remember is that two tangent circles and two parallel straight lines turn into each other if the center of inversion is the point of tangency of the two circles. We illustrate with the following problem.

Two circles ω_1 and ω_2 are tangent externally to each other at A. A common exterior tangent touches ω_1 at P and ω_2 at Q. The other common exterior tangent touches ω_1 at R and ω_2 at S. Prove that the circumcircles of triangles PAQ and RAS are tangent to each other.

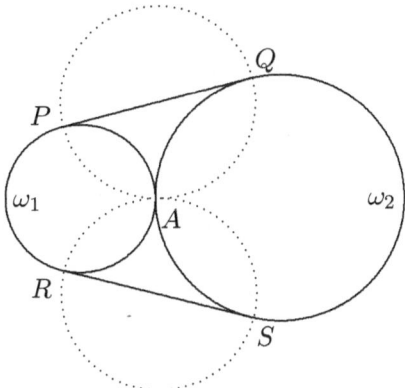

Figure 4.9

Invert with respect to A. Then ω_1 and ω_2 become parallel lines $P'R'$ and $Q'S'$. PQ and RS become circles ω_3' and ω_4', tangent to both $P'R'$ and $Q'S'$. The circumcircles of triangles PAQ and RAS become diameters $P'Q'$ and $R'S'$ of ω_3' and ω_4'. These diameters are orthogonal to PR and are therefore parallel to each other. Hence the two circumcircles are also tangent to each other.

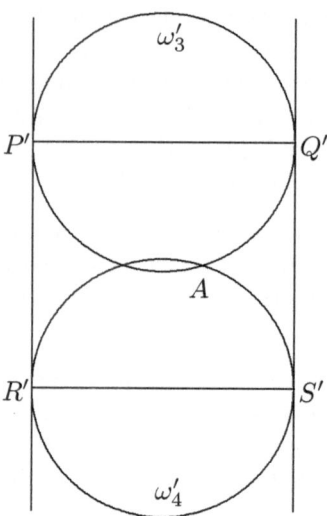

Figure 4.10

We give two more illustrative examples.

Four circles ω_1, ω_2, ω_3 and ω_4 are such that ω_1 and ω_2 intersect at A_1 and A_2, ω_2 and ω_3 intersect at B_1 and B_2, ω_3 and ω_4 intersect at C_1 and C_2, and ω_4 and ω_1 intersect at D_1 and D_2. Prove that if A_1, B_1, C_1 and D_1 are collinear or concyclic, then so are A_2, B_2, C_2 and D_2.

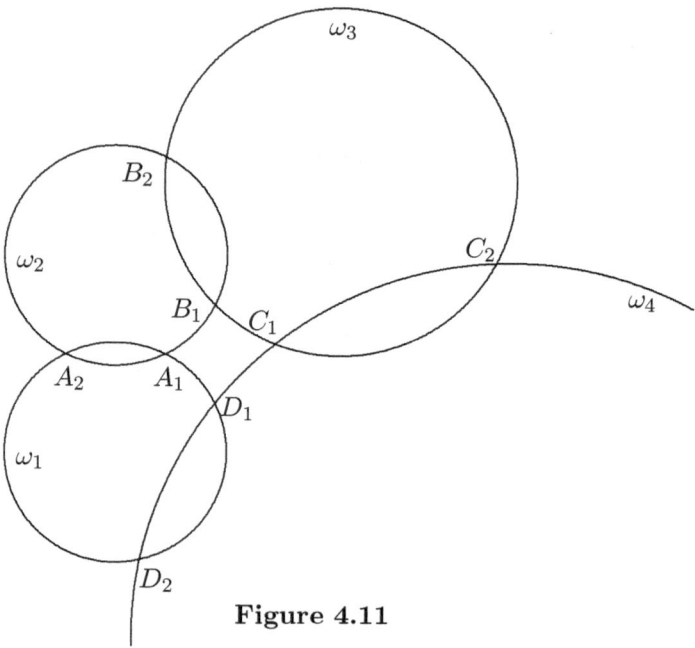

Figure 4.11

Here, we invert with respect to A_1. Then $A_1B_1C_1D_1$, ω_1 and ω_2 become the sides of triangle $A_2'B_1'D_1'$. Note that $B_2'B_1'C_1'C_2'$ is cyclic. Hence we have $\angle A_2'B_2'C_2' = \angle C_2'C_1'B_1'$. Similarly, $\angle A_2'D_2'C_2' = \angle C_2'C_1'D_1'$. Since A_1, B_1, C_1 and D_1 are either collinear or concyclic, C_1' lies on $B_1'D_1'$. It follows that $\angle C_2'C_1'B_1' + \angle C_2'C_1'D_1' = 180°$. Hence $\angle A_2'B_2'C_2' + \angle A_2'D_2'C_2' = 180°$, so that A_2', B_2', C_2' and D_2' are concyclic. Hence A_2, B_2, C_2 and D_2 are either collinear or concyclic.

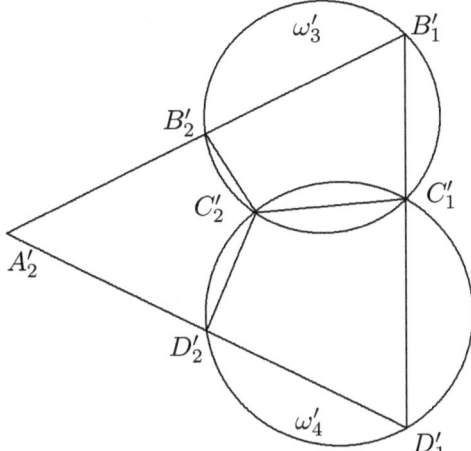

Figure 4.12

Three circles ω_1, ω_2 and ω_3 pass through O. A is the other point of intersection of ω_1 and ω_2, B is the other point of intersection of ω_2 and ω_3, and C is the other point of intersection of ω_3 and ω_1. The tangent to ω_2 at O intersects BC at D, the tangent at O to ω_3 intersects CA at E, and the tangent at O to ω_1 intersects AB at F. Prove that D, E and F are collinear.

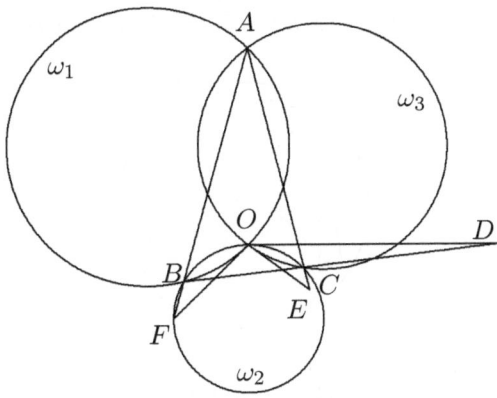

Figure 4.13

Invert with respect to O. Then the three circles turn into triangle $A'B'C'$ while the tangent lines OD, OE and OF turn into themselves. Hence OD', OE' and OF' are parallel to $B'C'$, $C'A'$ and $A'B'$ respectively. Moreover, D', E' and F' lie on the circumcircles of triangles $OB'C'$, $OC'A'$ and $OA'B'$ respectively. Let Q be the circumcenter of triangle $A'B'C'$. Then Q lies on the perpendicular bisectors of OD', OE' and OF'. Hence O, D', E' and F' are concyclic. It follows that D, E and F are collinear.

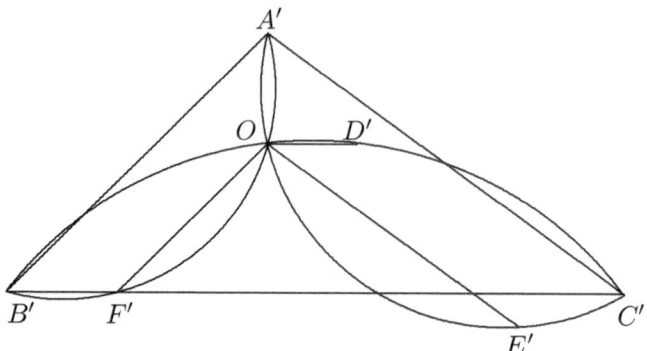

Figure 4.14

We now turn our attention to problems involving some famous results.

Simson's Theorem.
Let P be any point on the circumcircle of triangle ABC other than the vertices. From P, perpendiculars are dropped to BC, CA and AB, intersecting them at D, E and F respectively. Then D, E and F are collinear.

This can be proved without inversion as follows. Since $PBCA$ is cyclic, $\angle PAE = \angle PBC$, so that $\angle APE = \angle BPD$. Since $\angle PDB = 90° = \angle PFB$, $PBDF$ is cyclic. Similarly, $PEAF$ is also cyclic. It follows that we have $\angle EFA = \angle EPA = \angle BPD = \angle BFD$. Since A, F and B are collinear, D, E and F are also collinear.

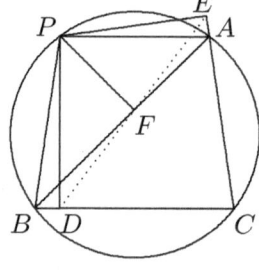

Figure 4.15

We now give a proof by inverting with respect to P. The circumcircle of ABC turns into the line $B'A'$ with C' on it. The side AB turns into the circumcircle of $PA'B'$, and PF, which is perpendicular to AB, becomes a diameter PF' of this circle. Similarly, PE' is a diameter of the circumcircle of $PC'A'$ and PD' is a diameter of the circumcircle of $PB'C'$. Since both $PBDF$ and $PFAE$ are both cyclic, F' lies on both $B'D'$ and $A'E'$. Since $\angle B'D'P = \angle B'C'P = \angle A'E'P$, $PD'F'E'$ is cyclic, so that D, E and F are collinear.

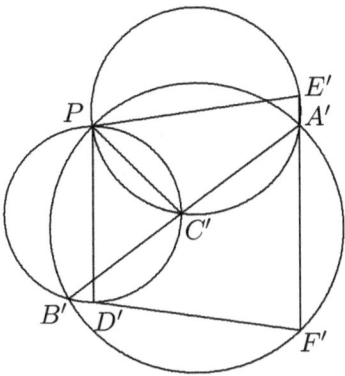

Figure 4.16

The next result is not so easy to prove without inversion.

Ptolemy's Inequality.
$AB \cdot CD + AD \cdot BC \geq AC \cdot BD$ for any convex quadrilateral $ABCD$, with equality if and only if the quadrilateral is cyclic.

Proof:
Invert with respect to A.

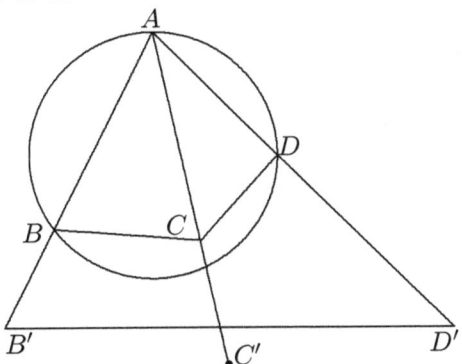

Figure 4.17

Now $B'D' = \frac{BD \cdot r^2}{AB \cdot AD}$, where r is the radius of inversion. Similarly, we have $B'C' = \frac{BC \cdot r^2}{AB \cdot AC}$ and $C'D' = \frac{CD \cdot r^2}{AC \cdot AD}$. It follows from the Triangle Inequality that $B'D' \le B'C' + C'D'$. Substituting in this the above expressions, we have $\frac{BD}{AB \cdot AD} \le \frac{BC}{AB \cdot AC} + \frac{CD}{AC \cdot AD}$, or $AC \cdot BD \le BC \cdot AD + CD \cdot AB$. Equality holds if and only if C' is collinear with B' and D'. Since the circumcircle of triangle BAD turns into the line $B'D'$, this holds if and only if C is concyclic with A, B and D.

Feuerbach's Theorem.

The circle which passes through the midpoints of the sides of a triangle is tangent to the triangle's incircle and excircles.

Proof:

We shall prove that the midpoint circle is tangent to the incircle and the excircle facing B. By symmetry, it will be tangent to the other two excircles. Let BC, CA and AB be tangent to the excircle facing B at K, L and M, and the incircle at N, P and Q, respectively. Now

$\qquad AB + AC - BC = (AQ + BQ) + (AN + CN) - (BP + CP) = 2AN$

and

$\qquad AB + AC - BC = (BK - AK) + (AM + CM) - (BL - CL) = 2CM$.

Hence $AN = CM$. Since E is the midpoint of AC, it is also the midpoint of E. Invert with respective to E and choose $EM = EN = \frac{BC - AB}{2}$ as the radius of inversion. Then both circles are orthogonal to the circle of inversion, and coincide with their respective images. Let XY be the other common interior tangent of these two circles, with X on AB and Y on BC.

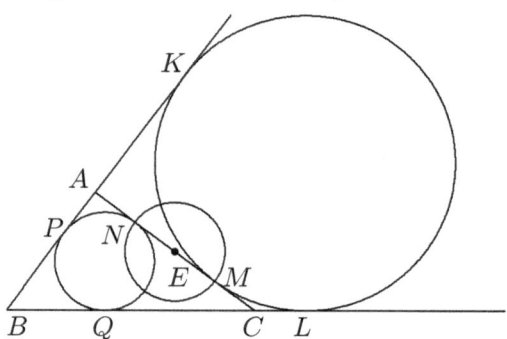

Figure 4.18

Let XY intersect DE at D' and EF at F'. If we can prove that D' and F' are the images of D and F respectively, then the midpoint circle inverts into the line XY, and the desired result follows. By symmetry, we have $BX = BC$ and $BY = BA$. Note that XFF' and XBY are similar triangles. It follows that $FF' = \frac{BY \cdot XF}{BX} = \frac{BA(BX-BF)}{BC} = \frac{BA(2BC-BA)}{2BC}$ and $EF' = EF - FF' = \frac{BC}{2} - \frac{2BC \cdot BA - BA^2}{2BC} = \frac{(BC-BA)^2}{2BC}$. Thus we have $EF \cdot EF' = \frac{BC}{2} \cdot \frac{(BC-BA)^2}{2BC} = (\frac{BC-BA}{2})^2$. Hence F' is indeed the inversive image of F. From the similar triangles XBY and $D'DY$, we can deduce in an analogous manner that D' is the inversive image of D.

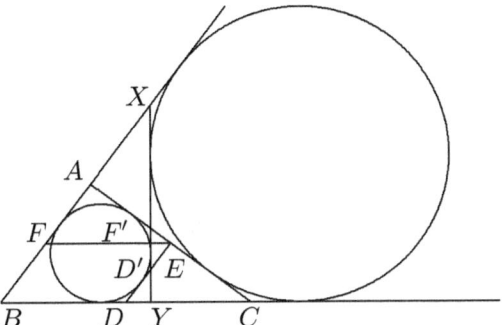

Figure 4.19

We conclude with a problem which uses inversion in combination with other transformations.

From a point O are four rays OA, OC, OB and OD in that order, such that $\angle AOB = \angle COD$. A circle tangent to OA and OB intersects a circle tangent to OC and OD at E and F. Prove that $\angle AOE = \angle DOF$.

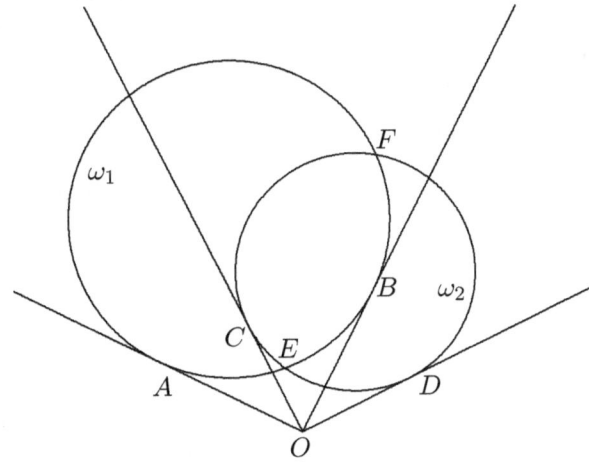

Figure 4.20

Let ω_1 be the circle tangent to OA at A and to OB at B. Let ω_2 be the circle tangent to OC at C and to OD at D. If ω_1 and ω_2 are of the same size, then both E and F will lie on the bisector of $\angle COB$, and the desired result follows immediately. Thus we may assume that ω_1 is larger than ω_2. Invert with respect to O so that A and B coincide with their respective images A' and B'. The rays OA, OC, OB and OD become the rays OA', OC', OB' and OD' respectively. The circle ω_1 coincides with its image ω_1' while the image of the circle ω_2 is another circle ω_2'. Note that the image E' of E is collinear with E and O, and the image F' of F is collinear with F and O.

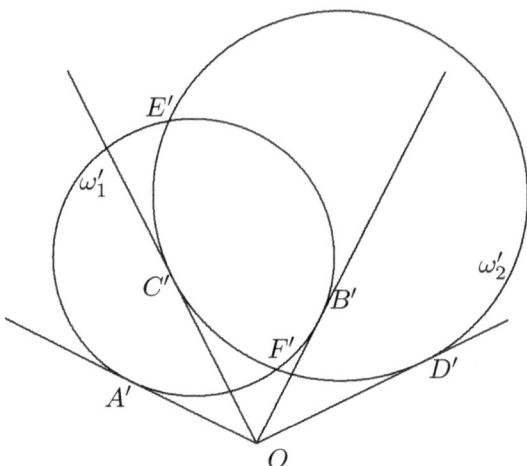

Figure 4.21

We now go from the first diagram to the second in a different way. First we perform a reflection about the bisector of $\angle COB$, and then a dilation from O so that D is mapped into A'. Note that the rays OA, OC, OB and OD become the rays OD', OB', OC' and OA' respectively, while the circle ω_2 becomes the circle ω_1'. By inversion, $OD \cdot OD' = OA^2$. Since D is mapped into A', A is mapped into D' so that the circle ω_1 becomes ω_2'. It follows that F is mapped into E' while E is mapped into F'. Thus the rays OE and OF become each other, and the desired result follows.

Exercises:

4. AB, AC and AD are three chords on a circle. Circles with AB and AC as diameters intersect at E, circles with AB and AD as diameters intersect at F, and circles with diameters AC and AD intersect at G. Prove that E, F and G are collinear.

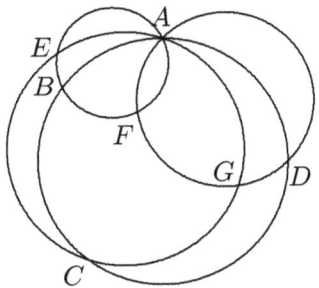

Figure 4.22

5. Four circles ω_1, ω_2, ω_3 and ω_4 are such that ω_1 and ω_2 touch at A, ω_2 and ω_3 touch at B, ω_3 and ω_4 touch at C and ω_4 and ω_1 touch at D. Prove that A, B, C and D are concyclic.

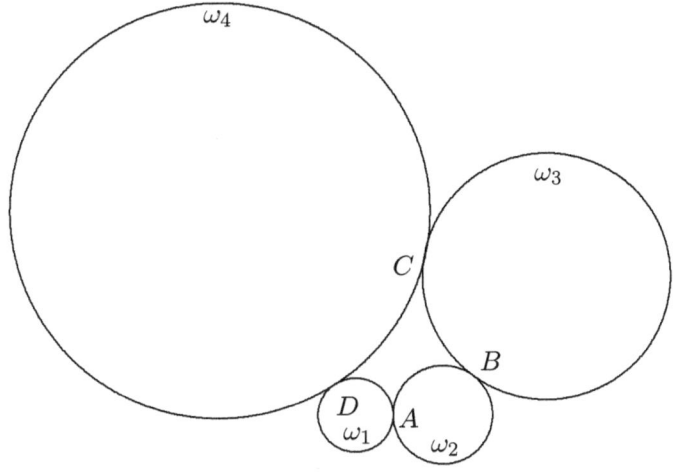

Figure 4.23

6. A, B and C are three points on a line and P is a point not on this line. Prove that the circumcenters of triangles PAB, PBC and PCA are concyclic with P.

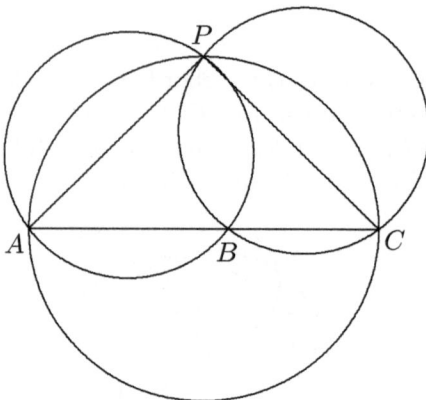

Figure 4.24

Section 3. Mohr-Mascheroni Constructions.

Constructions using a compass but without a straight edge are known as Mohr-Mascheroni constructions. Note that for convenience, we say a compass instead of a pair of compasses.

We should mention that there are three types of compasses. The two arms of a rusty compass cannot be adjusted, and thus the compass can only be used to draw circles of a fixed size. The two arms of a collapsing compass come together as soon as a circular arc has been drawn and the compass is moved off the paper. Thus it cannot be used to transfer distances. The modern compass has adjustable arms which remain in place until readjusted. For convenience, we shall be using the modern compass. It can be proved that anything constructible using a modern compass are also constructible using a rusty compass or a collapsing compass.

With only a compass, it is clearly not possible to draw any continuous portion of a line. However, in most construction problems, we only draw lines so that we can determine their points of intersection with other lines and circles. It may be possible to locate these points without drawing any lines.

To simplify our description of constructions, we use the notation $P(QR)$ to denote the circle with center P and radius QR. If R coincides with P, the notation is simplified to $P(Q)$.

We begin with two simple problems which do not involve inversion. Let A, B and P be non-collinear points. We wish to construct the point Q which is the image of P reflected across the line AB. This is accomplished by drawing the circles $A(P)$ and $B(P)$, intersecting each other again at Q.

Given points A and B, we wish to construct the point B_n on the extension of AB such that $AB_n = nAB$, where n is any non-negative integer. Clearly, $B_0 = A$ and $B_1 = B$. Draw the circles $B_0(B_1)$ and $B_1(B_0)$, cutting each other at C_1 and D_1. Draw the circle $C_1(D_1)$, cutting $B_1(B_0)$ at B_2. Draw the circle $B_2(B_1)$, cutting $B_1(B_0)$ at C_2 and D_2. Draw the circle $C_2(D_2)$, cutting $B_2(B_1)$ at B_3. This process can be continued until B_n is obtained. The justification is easy.

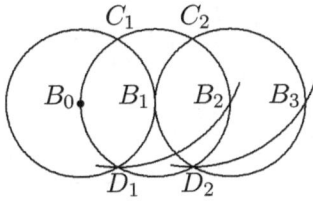

Figure 4.25

Given a circle ω of inversion with center O, we wish to construct the image A' of a point $A \neq O$. Draw the circle $A(O)$, cutting ω at C and D. Draw the circles $C(O)$ and $D(O)$, cutting each other at A'.

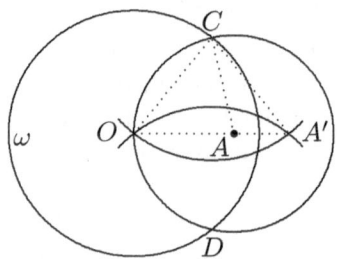

Figure 4.26

We now justify this construction. Since $AC = AO$, $\angle ACO = \angle AOC$. Since $CO = CA'$, $\angle AOC = \angle OA'C$. Hence triangles AOC and COA' are similar, so that $\frac{AO}{OC} = \frac{CO}{OA'}$. Since $OA \cdot OA' = OC^2$, A' is indeed the inversive image of A.

If A is too close to O, $A(O)$ will not cut ω in two points. We shall replace A by a point B such that $OB = nOA$ for some positive integer n, and $B(O)$ cuts $\mathcal{C}$ in two points. We can then construct its inverse B' with respect to ω. Then $OA' = nOB'$. We can obtain both B and A' from A and B' respectively.

Given a circle ω of inversion with center O, we wish to construct the circle which is the image of a line determined by two points A and B not passing through O. Construct the point C which is the image of O reflected across the line AB. Construct the image C' of C. The circle $C'(O)$ is the desired image.

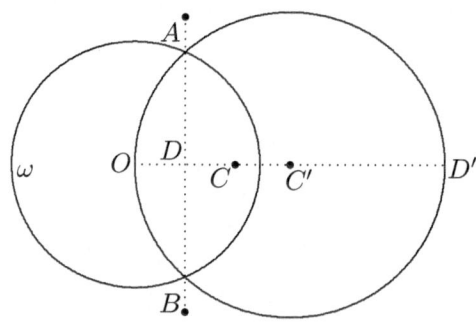

Figure 4.27

We now justify this construction. Let the line OC cut AB at D and $C'(O)$ again at D'. Since $OC = 2OD$, $OD' = 2OC'$ and C and C' are inverses, so are D and D'. Hence $C'(O)$ is indeed the image of AB.

Given a circle ω of inversion with center O, we wish to construct the circle which is the image of a circle $A(B)$ not passing through O. Construct the image C of O with respect to $A(B)$ and the respective images B' and C' of B and C with respect to ω. The circle $C'(B')$ is the desired image.

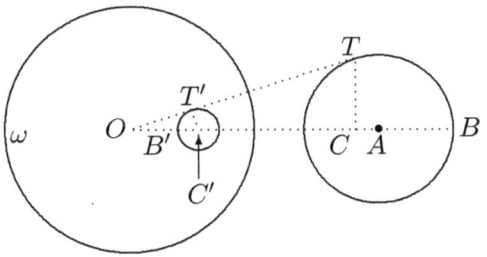

Figure 4.28

We now justify this construction. Let a tangent from O to $A(B)$ touch it at T and let T' be its inverse. Since C is the inverse of O with respect to $A(B)$, TC is perpendicular to OC. Hence $C'T'$ is perpendicular to OT'. It follows that $C'(B')$ is indeed the image of $A(B)$.

The **Mohr-Mascheroni Theorem** states that as long as only points are to be constructed, everything that can be done with a straight-edge and a compass can be done with a compass only. Choose a point O not lying on any construction lines and invert the whole diagram with respect to some circle ω with center O. Then all construction lines turn into circles and can be carried out with only a compass. When the desired point is obtained, we simply invert it back to the original diagram.

As an illustration, suppose we are to construct the midpoint C of a line segment AB. The usual Euclidean construction goes as follows. Draw the circles $A(B)$ and $B(A)$, cutting each other at D and E. Draw the line DE. cutting AB at C.

The corresponding Mohr-Mascheroni construction goes as follows. Invert the circles $A(B)$ and $B(A)$. Their points of intersection are the inverses D' and E' of D and E. Invert the lines AB and DE. Their points of intersection are O and the inverse C' of the desired point C. Finally, invert the point C' to obtain C.

Mohr-Mascheroni constructions do not necessarily involve the concept of inversion. For instance, the center of a given circle ω may be constructed as follows. Let A and B be any two points on the circle. Draw $A(B)$, intersecting ω again at C. Draw $B(A)$ and $C(A)$, intersecting each other at D. Draw $D(A)$, intersecting $A(B)$ at E and F. Draw $E(A)$ and $F(A)$, intersecting each other again at the center O of ω.

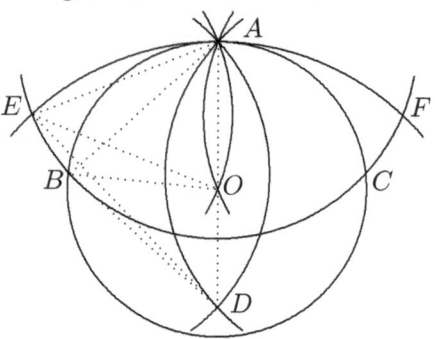

Figure 4.29

We now justify this construction. Note that the isosceles triangles DAE and EAO are similar since they have a common base angle. Hence $AO{\cdot}AD = AE^2 = AB^2$. It follows that triangles BAD and OAB are also similar. Since $BA = BD$, $OA = OB$. By symmetry about OA, O is indeed the desired center.

In the famous "Napoleon's Problem", a circle with its center is given. The task is to divide it into four equal arcs using a compass only. Let O be the center and A be any point on the circle. Draw $A(O)$, intersecting $O(A)$ at B. Draw $B(O)$, intersecting $O(B)$ at C. Draw $C(O)$, intersecting $O(C)$ at D. Draw $A(C)$ and $D(B)$, intersecting each other at E. Draw $A(OE)$, intersecting $A(O)$ at F and G. Then A, F, D and G are the desired division points.

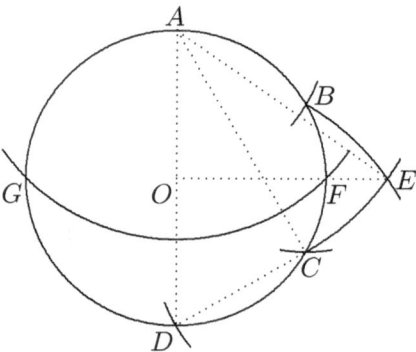

Figure 4.30

We now justify this construction. Take $OA = 1$. Then $CD = 1$ and $AD = 2$ so that $AE = AC = \sqrt{3}$. It follows that $AF = OE = \sqrt{2}$, so that $AFDG$ is indeed a square.

Exercises:

7. Given three non-collinear points, use a compass only to construct the center of the circle passing through them.

8. Find a solution to Napoleon's Problem drawing only five arcs.

9. (a) Given two adjacent vertices of a square, find the other two vertices using only a compass.

   (b) Given two opposite vertices of a square, find the other two vertices using only a compass.

Bibliography

[1] H. Eves, *A Survey of Geometry.* 2 vols. Allyn and Bacon, Boston, 2nd edition, 1972.

[2] Martin Gardner, *Mathematical Circus*, Mathematical Association of America, Washington (1995) 216–231.

[3] C. S. Ogilvy, *Excursions in Geometry*, Dover, Mineola, 1969.

Solutions to Exercises

1. The inversive images are shown in Figure 4.31.

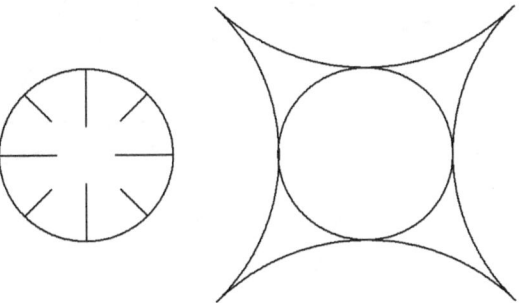

Figure 4.31

2. Let O be the center of the circle of inversion, and let a circle ω intersect it orthogonally at A and B. Let the line joiining O to the center of ω intersect ω at P and Q. Now the images of A and B are themselves. Since the two circles are orthogonal, OA is a tangent to ω, so that $OA^2 = OP \cdot OQ$. Hence Q is the inversive image of P. Now ω is determined by A, B and P. Hence its image ω' is determined by A, B and Q, which is ω itself.

3. Invert with respect to A. Then B', C' and D' become three collinear points. The variable circle through A and B becomes a variable line through B', which is tangent to a circle through C' and D' at T'. Now $(B'T')^2 = B'C' \cdot B'D'$ is constant. Hence the locus of T' is a circle with center B'. Depending on whether this circle passes through A or not, the locus of T is either a straight line or a circle.

4. Invert with respect to A. Then the original circle becomes the line $B'C'D'$. The other three circles become the lines $E'B'F'$, $E'C'G'$ and $F'G'D'$, and they are orthogonal to AB', AC, and AD, respectively. Hence $AE'B'C'$, $AB'F'D'$ and $AC'G'D'$ are cyclic quadrilaterals. It follows that we have $\angle B'E'C' = \angle B'AC'$, $\angle B'AF' = \angle B'D'F'$ and $\angle C'AG' = \angle C'D'G'$. Hence
$$\begin{aligned}
\angle F'AG' &= \angle B'AG' - \angle B'AF' \\
&= \angle B'AG' - \angle C'D'G' \\
&= \angle B'AG' - \angle B'D'F' \\
&= \angle B'AC' \\
&= \angle F'E'G'.
\end{aligned}$$

Hence A, E', F' and G' are concyclic, so that E, F and G are collinear.

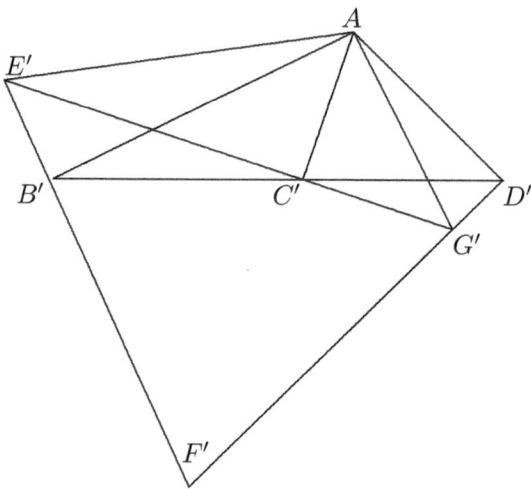

Figure 4.32

5. Invert with respect to A. Then ω_1 and ω_2 become a pair of parallel lines, tangent to ω'_4 and ω'_3 at D' and B' respectively. These two circles are tangent to each other at C'. Let P' and Q' be the centers of ω'_4 and ω'_3 respectively. Then C' lies on $P'Q'$. Since $C'D'$ and $C'B'$ are parallel, $\angle C'P'D' = \angle C'Q'B'$. Since $C'P =' D'P'$ and $C'Q' = B'Q'$,

$$\angle P'C'D' = \frac{1}{2}(180° - \angle C'P'D')$$
$$= \frac{1}{2}(180° - \angle C'Q'B')$$
$$= \angle Q'C'B'.$$

Hence C' also lies on $B'D'$, which means that A, B, C and D are concyclic.

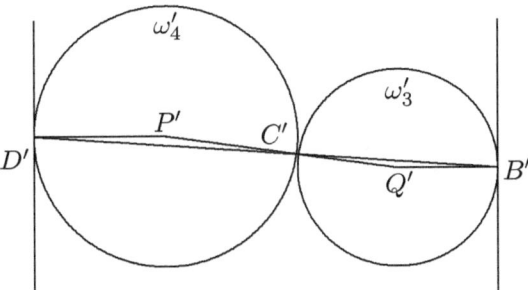

Figure 4.33

6. Let F, D and E be the respective circumcenters. Invert with respect to P. Then the circles become the sides of triangle $A'B'C'$. The images D', E' and F' of the circumcenters are the reflections of P across the respective sides.

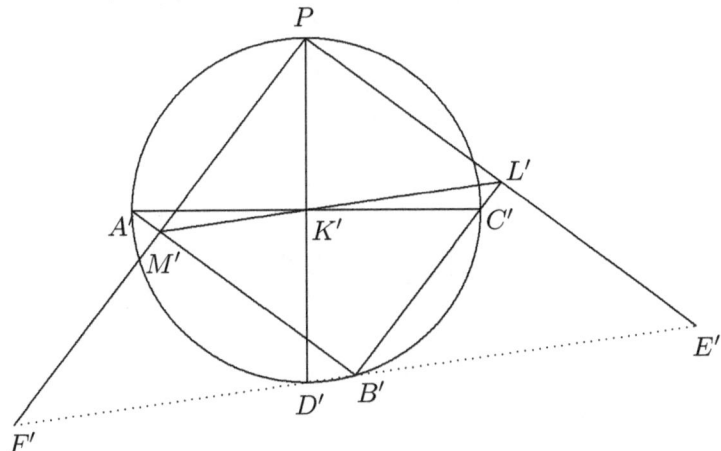

Figure 4.34

It follows that the midpoints K', L' and M' of PD', PE' and PF' are the feet of perpendiculars from P to the sides of triangle $A'B'C'$. Since A, B and C are collinear, P lies on the circumcircle of triangle $A'B'C'$. It follows that $M'K'L'$ is the Simson line of triangle $A'B'C'$, so that F', D' and E' are also collinear. Hence the circumcenters are concyclic with P.

7. Let the given points be A, B and C. Draw $A(B)$. Construct the inversive image C' of C with respect to $A(B)$. Reflect A across BC' to O', and construct the inversive image O of O' with respect to $A(B)$. Then O is the desired center.

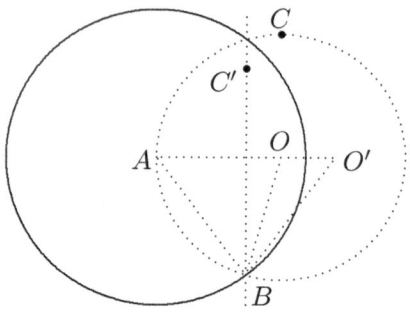

Figure 4.35

We now justify this construction. The circle through A, B and C is the inversive image of BC' with respect to $A(B)$. Since $AO \cdot AO' = AB^2$, triangles ABO and $AO'B$ are similar. Since $BA = BO'$, $OA = OB$. By symmetry about OA, O is indeed the desired center.

8. Let the given circle be $A(O)$.

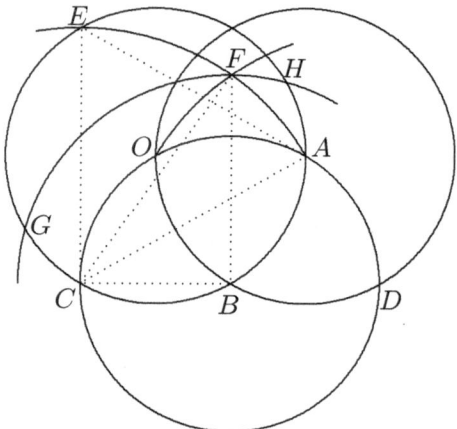

Figure 4.36

Draw $A(O)$, intersecting $O(A)$ at B. Draw $B(O)$, intersecting $O(A)$ again at C and $A(O)$ again at D. Draw $C(A)$ and $D(O)$, intersecting $O(A)$ again at E. Draw $D(O)$, intersecting $C(A)$ at F. Draw $B(F)$, intersecting $A(O)$ at G and H. Then B, G, E and H are the desired division points.

We now justify this construction. Take $OA=1$. Then $CF=AC=\sqrt{3}$. It follows that $BG = BH = BF = \sqrt{2}$. Since ACE is an equilateral triangle, $BGEH$ is indeed a square.

9. (a) Let the given vertices be A and B. Draw $A(B)$ and $B(A)$, intersecting each other at E and F. Draw $E(F)$, intersecting $B(A)$ at G. Draw $A(EF)$ and $G(EF)$, intersecting each other at H. Draw $B(H)$, intersecting $A(B)$ at D. Draw $A(BH)$, intersecting $B(A)$ at C. Then C and D are the desired vertices.

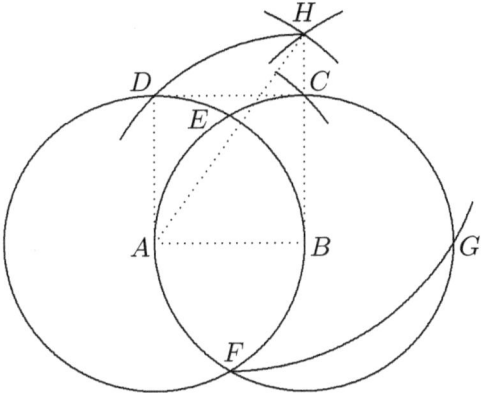

Figure 4.37

We now justify this construction. Take $AB = 1$. Then we have $AH = EF = \sqrt{3}$. It follows that $AC = BD = BH = \sqrt{2}$. Hence $ABCD$ is indeed a square.

(b) Let A and C be the given vertices.

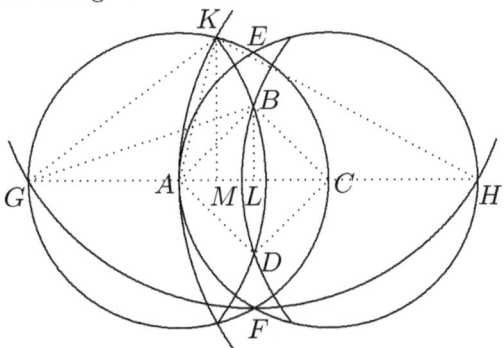

Figure 4.38

Draw $A(C)$ and $C(A)$, intersecting each other at E and F. Draw $E(F)$, intersecting $A(C)$ again at G and $C(A)$ again at H. Draw $H(A)$, intersecting $A(C)$ at K. Draw $E(K)$ and $F(EK)$, intersecting each other at B and D. Then B and D are the desired vertices.

We now justify this construction. Take $AC = 2$. Then $AK = 2$ and $AH = KH = 4$. With AK as base, we see that the area of triangle KAH is $\sqrt{15}$. With AH as base, the altitude $KM = \frac{\sqrt{15}}{2}$. Hence $HM = \frac{7}{2}$ and $GM = \frac{5}{2}$. It follows that $GB = GK = \sqrt{10}$. Let L be the foot of perpendicular from B to AC. Then $BL = 1$. Hence $ABCD$ is indeed a square.

Chapter Five: Convexity

Section 1. Figures

By a rectangle, we may mean one of several things. It may be a rectangular frame consisting only of four segments joined end to end at right angles. It may also be the region enclosed by this frame, along with all, none or part of the frame, as shown in Figure 5.1.

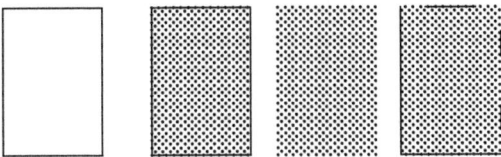

Figure 5.1

In Euclidean and other branches of classical geometry, we study figures with much regularity, such as polygons and circles. In this chapter, we study figures in general, often *without* regularity. We just scratch at the surface of this growing and fascinating branch of mathematics.

We begin with a fairly formal definition of a figure.

Let r be a positive number. The r-**neighborhood** of a point P in the plane is the set of all points in the plane at a distance less than r from P, and is denoted by $N(P, r)$. In other words, $N(P, r)$ is a disc with center P and radius r, without the circumference.

A point P in a set S is said to be an **interior point** of S if $N(P, r)$ is a subset of S for some r. A point P not in S is said to be an **exterior point** of S if $N(P, r)$ is disjoint from S for some r. A point P, whether in S or not, is said to be a **boundary point** of S if it is neither an interior point nor an exterior point of S.

In Figure 5.1, every point on the rectangular frame is a boundary points. All other points are exterior points. In each of the rectangular regions, every point inside the frame is an interior point, every point outside the frame is an exterior point, and every point of the frame is a boundary point.

A set is said to be **closed** if it contains all its boundary points. In Figure 5.1, only the rectangular frame and the first rectangular region are closed.

A set S is said to be **bounded** if S is subset of $N(P, r)$ for some point P and some positive number r. All four sets in Figure 5.1 are bounded. To see this, choose P to be the center of the rectangle and r to be greater than half the length of its diagonal.

© Springer International Publishing AG 2018
A. Liu, *S.M.A.R.T. Circle Minicourses*, Springer Texts in
Education, https://doi.org/10.1007/978-3-319-71743-2_5

A set S is said to be **simply connected** if for any two points A and B in S, there is a continuous path between A and B which lies entirely within S, and moreover, if any two such paths can be deformed into each other within S.

Let us make this clearer by putting a piece of thread over the first path. If we can push, pull or stretch the piece of thread so that it is over the second path, without passing over an exterior point of S, then the two continuous paths can be deformed into each other within S. The rectangular frame in Figure 5.1 is *not* simply connected, but the three rectangular regions are.

A point P in a simply connected set S is said to be an **articulation point** of S if there exist two points A and B in S such that every continuous path between A and B within S passes through P. None of the sets in Figure 5.1 has an articulation point. The "X-fighter" in Figure 5.2 is simply connected. It has an articulation point P since every continuous path with in S connecting a point in each "wing" must pass through P.

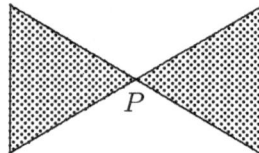

Figure 5.2

After all these preliminary definitions, we finally arrive at the crucial one. A set is said to be a **figure** if it is closed, bounded, simply connected and has no articulation points. In Figures 5.1 and 5.2, only the first rectangular region is a figure. Henceforth, this figure is meant when we use the term rectangle. The somewhat irregular curve in Figure 5.3 bounds a region which, together with the curve, is a figure. We leave off the shading for non-polygonal regions.

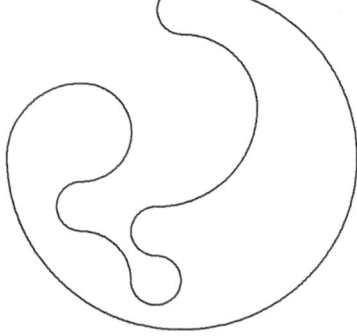

Figure 5.3

We now come to a group of results based on an important property of continuous functions, which we now state without proof.

The Intermediate Value Theorem.
Let $\mathcal{F}$ be a continuous function defined over the interval $[a, b]$ with $f(a) = A$, $f(b) = B$ and $A \neq B$. Then for any value C between A and B, there exists c in the interval (a, b) such that $f(c) = C$.

Theorem 1.
Given a figure and a direction, there exists a unique line in the given direction which bisects its area.

Proof:
We may take the direction to be horizontal. Since the figure is bounded, we can draw a horizontal line such that the entire figure is above it. The initial position of the line is said to be at height 0. Move the horizontal line continuously upward so that eventually the entire figure is below it, say at height 1. Consider the horizontal line at height t where $0 < t < 1$. Compute the area of the figure above the line and the area of the figure below the line, and define $f(t)$ to be the difference when the second sum is subtracted from the first. Now $f(0) > 0$ while $f(1) < 0$. By the Intermediate Value Theorem, there exists some t, $0 < t < 1$, such that $f(t) = 0$. Then the horizontal line at height t is the desired line.

Theorem 2.
Given a figure, there exist two perpendicular lines which quadrisect its area.

Proof:
By Theorem 1, for any direction, there is a line which bisects the area of the figure, and a second line perpendicular to the first which also bisects the area of the figure. Let the initial direction be horizontal and pointing to the right. Then it makes a $0°$ angle with the positive x-axis. If the area of the figure is quadrisected, there is nothing further to prove. If not, rotate the pair of perpendicular lines continuously (but not necessarily about a fixed point) so that it makes an angle θ with the positive x-axis and each bisecting the area of the figure. Define $f(\theta)$ to be the difference between the total area of a pair of opposite quadrants and the total area of the other pair of opposite quadrants. Since $f(0°) \neq 0$, $f(180°) \neq 0$ either since it is the same pair of perpendicular lines. On the other hand, the direction of the horizontal line is now pointing to the left, so that $f(180°)$ is opposite in sign to $f(0°)$. It follows from the Intermediate Value Theorem that $f(\theta) = 0$ for some θ between $0°$ and $180°$. The desired conclusion follows.

Theorem 3.
Given two figures, there exists a line which simultaneously bisects their areas.

Proof:

By Theorem 1, for any direction, there is a line which bisects the area of the first figure. Let the initial direction be horizontal and pointing to the right. Then it makes a $0°$ angle with the positive x-axis. If the area of the second figure is also bisected, there is nothing further to prove. If not, rotate the line continuously (but not necessarily about a fixed point) so that it makes an angle θ with the positive x-axis and bisecting the area of the first figure. Define $f(\theta)$ to be the difference when the area of the second figure on the left side of the line is subtracted from the area of the second figure on the right side of the line. Since $f(0°) \neq 0$, $f(180°) \neq 0$ either since it is the same line. On the other hand, the direction is now pointing to the left, so that $f(180°)$ is opposite in sign to $f(0°)$. It follows from the Intermediate Value Theorem that $f(\theta) = 0$ for some θ between $0°$ and $180°$. The desired conclusion follows.

The three-dimensional version of this result is known in the mathematical folklore as the Ham Sandwich Theorem. It states that if two slides of bread and a slab of ham, however crumbled, are hanging in mid air, then there is a plane which simultaneously bisects the volume of all three.

Note that in Theorems 1, 2 and 3, we can replace area with perimeter.

A line ℓ is said to be a **supporting line** of a figure $\mathcal{F}$ if ℓ contains at least one boundary point but no interior points of $\mathcal{F}$. A tangent to a circle is a supporting line of the circle. A line along a side of a square is a supporting line of the square.

A boundary point of a figure is said to be a **corner point** if at least two supporting lines of the figure pass through it. A boundary point is said to be a **regular point** if exactly one supporting line passes through it. A circle has no corner points. In a rectangle, only the vertices are corner points.

Infinitely many supporting lines of a figure pass through each corner point of it. In Figure 5.4, two of the infinitely many supporting lines through the corner point P of the figure are shown as dotted lines, bounded by the two supporting lines through P shown as solid lines.

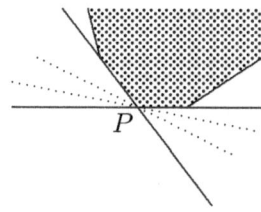

Figure 5.4

A polygon is said to be **circumscribed** about a figure if every side of the polygon is a supporting line of the figure.

Theorem 4.

For any figure, any direction and any integer $n \geq 3$, there exists a unique equiangular n-gon going counterclockwise, circumscribed about the figure and with one side in the given direction.

Proof:

For a fixed $n \geq 3$ and the direction of one side of an n-gon, the directions of the remaining sides are also determined. Since the figure is bounded, we can place it inside a large enough polygon whose sides are in the directions determined, and then contract the polygon until each side is stopped by the boundary of the figure. Then we have the desired circumscribed polygon, and it is unique.

Theorem 5.

For any figure, there exists a square circumscribed about it.

Proof:

By Theorem 4, there exists a unique rectangle $A_\theta B_\theta C_\theta D_\theta$, going counterclockwise, circumscribed about the figure and with $C_\theta D_\theta$ making an angle θ with the positive x-axis. Define $f(\theta) = A_\theta B_{theta} - B_\theta C_\theta$. Now $A_{90^\circ} = B_{0^\circ}$, $B_{90^\circ} = C_{0^\circ}$, $C_{90^\circ} = D_{0^\circ}$ and $D_{90^\circ} = A_{0^\circ}$. Hence $f(90^\circ) = -f(0^\circ)$. If $f(0^\circ) = 0$, then $A_{0^\circ} B_{0^\circ} C_{0^\circ} D_{0^\circ}$ is a square circumscribed about the figure. If not, we have $f(0^\circ) < 0 < f(90^\circ)$ or $f(0^\circ) > 0 > f(90^\circ)$. In either case, $f(\theta) = 0$ for some θ, $0^\circ < \theta < 90^\circ$, and $A_\theta B_\theta C_\theta D_\theta$ is a square circumscribed about the figure.

Theorem 6.

For any figure, there exists an equiangular hexagon circumscribed about it and having opposite sides equal.

Proof:

By Theorem 4, there exists a unique equiangular hexagon $A_\theta B_\theta C_\theta D_\theta E_\theta F_\theta$, going counterclockwise, circumscribed about the figure and with $E_\theta F_\theta$ making an angle θ with the positive x-axis. Define $f(\theta) = A_\theta B_\theta - D_\theta E_\theta$. Note that we have $A_{180^\circ} = D_{0^\circ}$, $B_{180^\circ} = E_{0^\circ}$, $C_{180^\circ} = F_{0^\circ}$, $D_{180^\circ} = A_{0^\circ}$, $E_{180^\circ} = B_{0^\circ}$ and $F_{180^\circ} = C_{0^\circ}$. Hence $f(180^\circ) = -f(0^\circ)$. If $f(0^\circ) = 0$, then $A_{0^\circ} B_{0^\circ} C_{0^\circ} D_{0^\circ} E_{0^\circ} F_{0^\circ}$ is a hexagon with the desired properties. If not, we have $f(0^\circ) < 0 < f(180^\circ)$ or $f(0^\circ) > 0 > f(180^\circ)$. In either case, $f(\theta) = 0$ for some θ, $0^\circ < \theta < 180^\circ$ by the Intermediate Value Theorem, and $A_\theta B_\theta C_\theta D_\theta E_\theta F_\theta$ is a hexagon with the desired properties.

Theorem 7.

For any figure, there exists an equiangular hexagon circumscribed about it and having an axis of symmetry.

Proof:

By Theorem 4, there exists a unique equilateral triangle $A_\theta B_\theta C_\theta$ going counterclockwise, circumscribed about the figure and with $B_\theta C_\theta$ making an angle θ with the positive x-axis. Theorem 4 also guarantees the existence of a unique equilateral triangle $D_\theta E_\theta F_\theta$ going counterclockwise, circumscribed about the figure and with $E_\theta F_\theta$ opposite in direction to $B_\theta C_\theta$. Let O_θ be the center of $A_\theta B_\theta C_\theta$, and P_θ, Q_θ and R_θ be the intersections of $O_\theta A_\theta$ with $E_\theta F_\theta$, $O_\theta B_\theta$ with $F_\theta D_\theta$ and $O_\theta C_\theta$ with $D_\theta E_\theta$, respectively. Define $f(t) = E_\theta P_\theta - F_\theta P_\theta$. If $f(0°) = 0$, then the intersection of $A_{0°} B_{0°} C_{0°}$ with $D_{0°} E_{0°} F_{0°}$ is a hexagon with the desired properties. Otherwise, note that $E_\theta P_\theta^2 + F_\theta Q_\theta^2 + D_\theta R_\theta^2 = F_\theta P_\theta^2 + D_\theta Q_\theta^2 + E_\theta R_\theta^2$. Without loss of generality, we may assume that $E_\theta P_\theta < F_\theta P_\theta$ while $F_\theta Q_\theta > D_\theta Q_\theta$. Now $A_{120°} = B_{0°}$, $B_{120°} = C_{0°}$, $C_{120°} = A_{0°}$, $D_{120°} = E_{0°}$, $E_{120°} = F_{0°}$, $F_{120°} = D_{0°}$, $P_{120°} = Q_{0°}$, $Q_{120°} = R_{0°}$ and $R_{120°} = P_{0°}$. It follows that $f(120°) = -f(0°)$ and $f(0°) < 0 < f(120°)$. By the Intermediate Value Theorem, $f(\theta) = 0$ for some θ, $0° < \theta < 120°$, and the intersection of $A_\theta B_\theta C_\theta$ with $D_\theta E_\theta F_\theta$ is a hexagon with the desired properties.

Let a direction be given. Since a figure is bounded, we can put it between a pair of lines in that direction. By moving the lines towards each other until stopped by the boundary of the figure, we have a pair of supporting lines of the figure in that direction, and it is unique.

The distance between the pair of supporting lines of a figure in a given direction is called the **width** of the figure in that direction. The minimum width of a figure is called its **breadth**, and the maximum width is called its **diameter**.

Theorem 8.

If a pair of supporting lines of a figure defines its diameter, then each contains exactly one point of the figure, and the line joining these two points is perpendicular to the supporting lines.

Proof:

Let ℓ and m be a pair of supporting lines at the maximum distance apart. Let A be a point on ℓ and B be a point on m. Then AB must be perpendicular to the supporting lines. Otherwise, the distance between ℓ and m is less than AB, and therefore less than the distance between the supporting lines perpendicular to AB. Suppose ℓ contains another boundary point A' of the figure. Then $A'B$ cannot also be perpendicular to ℓ and m, and we have the same contradiction as before.

Theorem 9.

The diameter of a figure is equal to the maximum distance between two of its points.

Proof:

Let ℓ and m be a pair of supporting lines at the maximum distance d apart. Let A be a point on ℓ and B be a point on m. By Theorem 8, AB is perpendicular to the supporting lines. Hence $AB = d$. Suppose there exist two other points C and D of the figure such that $CD > d$. Consider the pair of supporting lines perpendicular to CD. Then their distance apart is greater than d, which is a contradiction.

Theorem 9 yields an equivalent definition of the diameter of a figure.

Theorem 10.

If two points A and B of a figure are at the maximum distance apart, then the two lines passing through A and B respectively and perpendicular to AB are supporting lines of the figure.

Proof:

Draw a pair of supporting lines of the figure perpendicular to AB. Then the entire segment AB lies between these two lines. By Theorem 9, the maximum distance between a pair of supporting lines of the figure is equal to AB. Hence each of A and B lies on one of the two lines.

The smallest circle that can cover a figure is called the **circumcircle** of the figure. The center and the radius of the circumcircle are called the **circumcenter** and the **circumradius** respectively. Since a figure is bounded, we can place it inside a large enough circle, and then contract the circle until stopped by the boundary of the figure. Thus the figure has a circumcircle, and it is unique. The circumcircle of a figure either contains two points that are diametrically opposite or three points that form an acute triangle.

Note that the definition of circumcircle here is different from the definition of the Euclidean circumcircle of a polygon. Not every polygon has an Euclidean circumcircle. For an obtuse triangle, the circumcircle has the longest side as its diameter, and it is considerably smaller than the Euclidean circumcircle. The two definitions coincide if the triangle is not obtuse.

Jung's Theorem.

If R is the circumradius of a figure of diameter 1, then $\frac{1}{2} \le R \le \frac{1}{\sqrt{3}}$.

Proof:

The greatest distance between two points of the figure is at most $2R$. Hence $2R \ge 1$ or $R \ge \frac{1}{2}$. This is attained when the figure is a circle of radius $\frac{1}{2}$. We now prove the upper bound. If the circumcircle contains two diametrically opposite points A and B of the figure, then $2R = AB \le 1$ and $R \le \frac{1}{2} < \frac{1}{\sqrt{3}}$. Otherwise, it contains three points of the figure that form an acute triangle ABC. One of the angles, say C, is at least $60°$. By the Sine Law, we have $2R = \frac{AB}{\sin C} \le \frac{1}{\sin 60°} = \frac{2}{\sqrt{3}}$. Hence $R \le \frac{1}{\sqrt{3}}$, and this is attained when the figure is an equilateral triangle of side 1.

Corollary.
Every figure of diameter 1 can be covered by a circle of radius $\frac{1}{\sqrt{3}}$, and this is best possible.

Pal's First Theorem.
Every figure of diameter 1 can be covered by an equilateral triangle of side $\sqrt{3}$, and this is best possible.

Proof:
That this is best possible can be seen if we take the figure to be a circle of radius $\frac{1}{2}$. Now take an arbitrary direction. By Theorem 4, there exists a unique equilateral triangle ABC going counterclockwise, circumscribed about the figure and with BC in the given direction. Theorem 4 also guarantees the existence of a unique equilateral triangle DEF going counterclockwise circumscribed about the figure and with EF opposite in direction to BC. Take an arbitrary point O in the figure. Denote its distances to BC, CA, AB, EF, FD and DE by u, v, w, x, y and z respectively. Then $u + x \leq 1$, $v + y \leq 1$ and $w + z \leq 1$ since the figure has diameter 1. Hence either $u + v + w \leq \frac{3}{2}$ or $x + y + z \leq \frac{3}{2}$. Without loss of generality, we may assume that $u + v + w \leq \frac{3}{2}$. Now the altitude of ABC is equal to $u + v + w$. Hence its side is less than or equal to $\sqrt{3}$. Since ABC covers the figure, an equilateral triangle of side $\sqrt{3}$ certainly can also.

Pal's Second Theorem.
Every figure of diameter 1 can be covered by a regular hexagon of side $\frac{1}{\sqrt{3}}$, and this is best possible.

Proof:
That this is best possible can be seen if we take the figure to be a circle of radius $\frac{1}{2}$. By Theorem 6, there exists an equiangular hexagon $ABCDEF$ circumscribed about the figure and having opposite sides equal. Now the distance between AB and DE is at most 1 as the figure has diameter 1. If the distance is less than 1, we displace AB and DE away from each other by the same amount so that the distance between them is 1. The same process is repeated on the other pairs of opposite sides if necessary, so that each pair of opposite sides of the enlarged hexagon is at a distance 1 apart. Since $ABCDEF$ has opposite sides equal, so does the new hexagon. It is easy to verify that it is a regular hexagon of side $\frac{1}{\sqrt{3}}$.

Borsuk's Theorem.
Every figure of diameter 1 can be partitioned into three figures each of diameter less than 1.

Proof:

By Pal's Second Theorem, the figure can be covered by a regular hexagon $ABCDEF$ of side $\frac{1}{\sqrt{3}}$. Let O be the center of the hexagon, and P, Q and R be the midpoints of AB, CD and EF respectively. Then OP, OQ and OR partition the hexagon into three congruent figures. It is easy to see that the diameter of each is equal to $PQ = \frac{\sqrt{3}}{2} < 1$. This partition induces one for the figure as desired.

Exercises

1. For any figure, prove that there exists a line which simultaneously bisects its area and perimeter.

2. For any figure, prove that there exists an equiangular pentagon circumscribed about it and having an axis of symmetry.

3. Prove that every figure of diameter 1 can be covered by a square of side 1, and this is best possible.

Section 2. Convex Figures

A set S is said to be **convex** if for any two points P and Q in S, the segment PQ is in S. In Figure 5.1, the rectangular frame and the last rectangular region are not convex, but the other two rectangular regions are.

Convex subsets of a line are the entire line, every closed ray (with the endpoint), every open ray (without the endpoint), every closed segment (with both endpoints), every half-open half-closed segment (with only one endpoint), every open segment (without endpoints), as well as a single point. There are no others. Some convex subsets of the plane are the entire plane and every closed half-plane (with the boundary).

For any two convex sets S_1 and S_2, their intersection $S = S_1 \cap S_2$ must be convex. Consider any two points P and Q in S. Then both are in S_1. Since S_1 is convex, the segment PQ is in S_1. Similarly, the segment PQ is in S_2. Hence the segment PQ is in S, which is therefore convex.

A **convex figure** is a figure that is convex. Convex figures contained in a line are closed segments as well as single points.

A **convex polygon** is a polygonal figure that is convex. A line along a side of a convex polygon is a supporting line of the polygon, and the polygon itself is the intersection of finitely many closed half-planes.

Here are some of the basic properties of a convex figure $\mathcal{F}$.

(1) For any interior points P and Q of $\mathcal{F}$, every point on the segment PQ is an interior point of $\mathcal{F}$.

(2) For any interior point P and any boundary point Q of $\mathcal{F}$, every point on the segment PQ except Q is an interior point of $\mathcal{F}$.

(3) For any boundary points P and Q of $\mathcal{F}$, either every point on the segment PQ is a boundary point of $\mathcal{F}$ or every point on the segment PQ except P and Q is an interior point of $\mathcal{F}$.

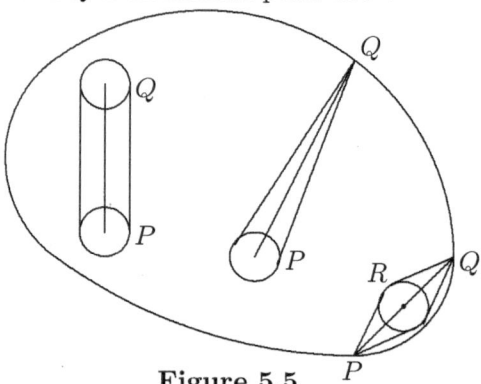

Figure 5.5

A convex figure $\mathcal{F}$ is shown in Figure 5.5. On the left, both P and Q are interior points of $\mathcal{F}$. Hence each has a neighborhood contained entirely in $\mathcal{F}$. Since $\mathcal{F}$ is convex, the entire region bounded by the two lines and two semicircles is in $\mathcal{F}$. Now any point on the segment PQ has a neighborhood which lies within this region. Hence it is an interior point. In the middle of Figure 5.5, Q is a boundary point and P is an interior point. The entire region bounded by the two lines and the circular arc is in $\mathcal{F}$, and any point on the segment PQ has a neighborhood which lies within this region.

Finally, let both P and Q be boundary points of $\mathcal{F}$. If there is at least one interior point R of $\mathcal{F}$ on the segment PQ, as shown in Figure 5.5 on the right, then the entire region bounded by the four lines and the two circular arcs is in $\mathcal{F}$. If the segment PQ does not contain any interior point of $\mathcal{F}$, then obviously every point on it is a boundary point of $\mathcal{F}$.

Conversely, any figure with property (1), (2) or (3) must be convex.

Theorem 11.
A figure $\mathcal{F}$ is convex if and only if every boundary point of $\mathcal{F}$ is either a corner point or a regular point.

Proof:
Suppose $\mathcal{F}$ has a boundary point P with no supporting lines passing through it. Consider all rays from P which pass through at least one other point of $\mathcal{F}$. The union of these rays is an angle greater thant $180°$ It is easy to see that there are two points in $\mathcal{F}$ such that the segment between them contains at least one exterior point. Hence $\mathcal{F}$ is not convex. Conversely, suppose $\mathcal{F}$ is not convex. Then there exist two points P and Q of $\mathcal{F}$ such that PQ contains an exterior point R of $\mathcal{F}$. We may assume that P is an interior point of $\mathcal{F}$, so that there is a boundary point S of $\mathcal{F}$ between P and R. The line PQ is not a supporting line of $\mathcal{F}$ since it passes through an interior point P. No other line through S can be a supporting line of $\mathcal{F}$ since P and Q will be on opposite sides of it. Hence S is a boundary point with no supporting lines passing through it.

Theorem 12.
A figure $\mathcal{F}$ is convex if and only if every line containing at least one interior point of $\mathcal{F}$ contains exactly two boundary points of $\mathcal{F}$.

Proof:
Suppose $\mathcal{F}$ is convex. Let P be an interior point of $\mathcal{F}$ and let ℓ be a line passing through P. Since ℓ is also convex, its intersection with $\mathcal{F}$ is convex, in fact, a convex figure. It cannot be a point since some neighborhood of P is contained in $\mathcal{F}$. Hence it is a closed interval. Clearly, the two endpoints are boundary points of $\mathcal{F}$. By Basic Property (3), since the interval contains an interior point P, the only boundary points are the two endpoints.

Conversely, suppose $\mathcal{F}$ is not convex. Then there exist two points P and Q of $\mathcal{F}$ such that PQ contains an exterior point R of $\mathcal{F}$. We may assume that P is an interior point of $\mathcal{F}$. Now there is at least one boundary point of $\mathcal{F}$ between Q and R (which may be Q itself), at least one boundary point between P and R, and at least one beyond RP since $\mathcal{F}$ is bounded. Hence the line through P and Q contains an interior point and at least three boundary points of $\mathcal{F}$.

Theorem 13.
If each of two segments in a convex figure $\mathcal{F}$ is equal in length to the diameter of $\mathcal{F}$, then they must intersect in $\mathcal{F}$.

Proof:
Let the segments be AB and CD. Suppose they do not intersect in $\mathcal{F}$. Then they form opposite sides of a convex quadrilateral, say $ABCD$. Let E be the point of intersection of AC and BD. By the Triangle Inequality, $AC + BD = (AE + BE) + (CE + DE) > AB + CD = 2AB$. Hence one of AC and BD must be longer than AB, contradicting the assumption that AB is a diameter.

Theorem 14.
If two convex figures are disjoint, then there is a line disjoint from both and with one figure on each side of it.

Proof:
Let A and B be points in the two figures respectively such that the distance AB is minimum. We claim that the line through A perpendicular to AB is a support line of the first figure. Suppose to the contrary that there is a point P of the first figure which lies on the same side of this line as B. Let Q be the foot of perpendicular from B to the line AP. If Q lies between A and P or coincides with P, then Q also belongs to the first figure. Now $BQ < AB$, and this is a contradiction. If Q lies beyond P, then $\angle APB > 90°$ so that $AB > BP$. Again we have a contradiction. Thus our claim is justified. By symmetry, the line through B perpendicular to AB is a support line of the second figure, and any line perpendicular to AB and lying between these two support lines separates the two convex figures.

Corollary.
If a convex figure is disjoint from a closed half-plane, then there is a line disjoint from both and with the figure on the opposite side of the line to the half-plane.

The **convex hull** of a set $\mathcal{S}$ is the smallest convex set containing $\mathcal{S}$, and is denoted by $H(\mathcal{S})$. Clearly, $H(\mathcal{S})$ is the intersection of all convex sets containing $\mathcal{S}$. Moreover, a figure is convex if and only if it is equal to its convex hull.

Theorem 15.
Every supporting line of a figure $\mathcal{F}$ is a supporting line of $H(\mathcal{F})$, and conversely.

Proof:
Suppose ℓ is a supporting line of $\mathcal{F}$. Then one of the half-planes defined by ℓ contains $\mathcal{F}$. Hence it also contains $H(\mathcal{F})$. Since ℓ passes through a point of $\mathcal{F}$, it also passes through a point of $H(\mathcal{F})$. Hence it is a supporting line of $H(\mathcal{F})$. Conversely, suppose ℓ is a supporting line of $H(\mathcal{F})$. If it is disjoint from $\mathcal{F}$, then by the Corollary to Theorem 14, there is a line m which separates ℓ from $\mathcal{F}$. Thus m defines a half-plane which contains $H(\mathcal{F})$, so that ℓ could not have been a supporting line of $H(\mathcal{F})$. It follows that ℓ contains a point P of $\mathcal{F}$. If P is an interior point of $\mathcal{F}$, it will also be an interior point of $H(\mathcal{F})$. This contradicts the assumption that ℓ is a supporting line of $H(\mathcal{F})$. Hence ℓ contains at least one boundary point of $\mathcal{F}$ but no interior points of $\mathcal{F}$, so that it is a supporting line of $\mathcal{F}$.

Corollary.

(a) Every corner point of a figure $\mathcal{F}$ is a corner point of $H(\mathcal{F})$, and every regular point of $\mathcal{F}$ is a regular point of $H(\mathcal{F})$.

(b) The diameter and breadth of a figure $\mathcal{F}$ is equal to the diameter and breadth of $H(\mathcal{F})$ respectively.

Theorem 16.
If $P_1 P_2 \ldots P_n$ is a convex n-gon contained in a figure $\mathcal{F}$, then the perimeter of the n-gon is less than or equal to that of $\mathcal{F}$, with equality if and only if the n-gon is $\mathcal{F}$ itself.

Proof:
Let the extension of $P_1 P_2$ intersect the boundary of $\mathcal{F}$ for the first time at Q_1. Let Q_2, Q_3, $\ldots$, Q_n be defined similarly by $P_2 P_3$, $P_3 P_4$, $\ldots, P_n P_1$ respectively. Now $P_1 Q_n + (Q_n Q_1) \geq P_1 Q_1$, where $(Q_n Q_1))$ denotes the length of the part of the boundary of $\mathcal{F}$ between Q_n and Q_1. Similarly, we have $P_2 Q_1 + (Q_1 Q_2) \geq P_2 Q_2$, $\ldots$, $P_n Q_{n-1} + (Q_{n-1} Q_n) \geq P_n Q_n$. Adding these inequalities yields the desired result.

Corollary.

(a) The perimeter of a convex figure contained in a figure $\mathcal{F}$ is less than or equal to the perimeter of $\mathcal{F}$, with equality if and only if the convex figure is $\mathcal{F}$ itself.

(b) The perimeter of a figure $\mathcal{F}$ is greater than or equal to the perimeter of $H(\mathcal{F})$, with equality if and only if $\mathcal{F}$ is convex.

A polygon $P_1 P_2 \ldots P_n$ is said to be **inscribed** in a figure $\mathcal{F}$ if P_k is a boundary point of $\mathcal{F}$ for $1 \le k \le n$, with P_1, P_2, $\ldots$, P_n appearing in that order along the boundary of $\mathcal{F}$.

Theorem 17.
A figure $\mathcal{F}$ is convex if and only if every polygon inscribed in $\mathcal{F}$ is convex.

Proof:
Suppose $\mathcal{F}$ is convex and $P_1 P_2 \ldots P_n$ is an n-gon inscribed in $\mathcal{F}$. Then P_1 and P_2 divide the boundary of $\mathcal{F}$ into two arcs, one of which contains all of P_3, P_4, $\ldots$, P_n. This arc plus the segment $P_1 P_2$ is the boundary of a convex figure since this figure is the intersection of $\mathcal{F}$ and the half-plane containing the arc. It is impossible for the other half-plane to contain any point of the n-gon without containing at least one vertex of it. It follows that $P_1 P_2$ is a supporting line of the n-gon. The same conclusion holds for each side of the n-gon, It follows easily that the n-gon is convex. Conversely, suppose $\mathcal{F}$ is not convex. Then there exist two points P and Q of $\mathcal{F}$ such that PQ contains an exterior point R of $\mathcal{F}$. Let S be the boundary point of $\mathcal{F}$ between P and R closest to R and T be the boundary point of $\mathcal{F}$ between Q and R closest to R, as shown in Figure 5.6. Let U be any point on the boundary of $\mathcal{F}$ between S and T, and V be any boundary point of $\mathcal{F}$ beyond RU. Then the quadrilateral $SUTV$ is not convex.

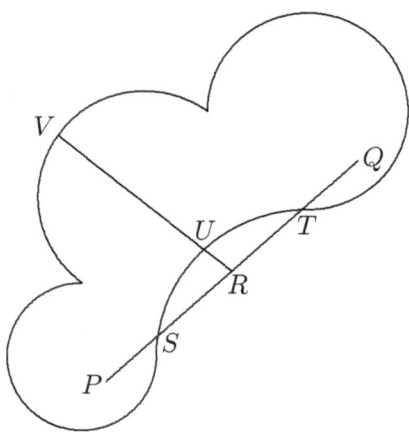

Figure 5.6

A point of a figure is said to be an **extreme** point if it does not lie on a segment determined by two other points of the figure. It is easy to see that every extreme point must be a boundary point. A figure is said to be **strictly** convex if every boundary point of it is an extreme point.

Theorem 18.

Given a strictly convex figure and a direction, there exists a unique hexagon $ABCDEF$ inscribed in the figure such that AB, FC and ED are all in the given direction, AB and DE are equidistant from FC, and $AB = DE = \frac{1}{2}FC$.

Proof:

Let A_dB_d be a chord of the figure of length d in the given direction. Let E_dD_d be a chord of equal length and in the same direction as A_dB_d. Let F_dC_d be the chord equidistant from A_dB_d and E_dD_d. Let d vary from 0 to the maximum length m of a chord in the same direction. Define $f(d) = F_dC_d - 2A_dB_d$. Note that $f(0) > 0$ while $f(m) = -m$. By the Intermediate Value Theorem, $f(d) = 0$ for some d in $(0, m)$, and in this position, the hexagon $ABCDEF = A_dB_dC_dD_dE_dF_d$ has the desired property. We now prove uniqueness. Suppose $A'B'C'D'E'F'$ is another such hexagon. We may assume that $CF \geq C'F'$. Let h denote the distance between CF and $C'F'$. It is easy to see that if h=O, then the two hexagons are identical. Now let $C'F'$ be on the same side of CF as AB. Then the distance between $A'B'$ and AB is at least $2h$. Let $A'F$ intersect AB at M and $C'F'$ at P. Let $B'C$ intersect AB at N and $C'F'$ at Q. Then $MN - A'B' \geq 2(CF - PQ)$. It now follows that $CF - PQ > CF - C'F' = 2(AB - A'B') > 2(MN - A'B') \geq 4(CF - PQ)$. This is clearly impossible.

Theorem 19.

In any strictly convex figure, a hexagon $ABCDEF$ can be inscribed such that AB, DE and CF are parallel, BC, EF and AD are parallel, CD, FA and BE are parallel, and AD, BE and CF are concurrent.

Proof:

By Theorem 18, we can inscribe in the figure a hexagon $A_\theta B_\theta C_\theta D_\theta E_\theta F_\theta$ with $E_\theta F_\theta$ making an angle θ with the positive x-axis, such that $A_\theta B_\theta$, $D_\theta E_\theta$ and $C_\theta F_\theta$ are parallel, $A_\theta B_\theta$ and $D_\theta E_\theta$ are equidistant from $C_\theta F_\theta$, and $A_\theta B_\theta = D_\theta E_\theta = \frac{1}{2}C_\theta F_\theta$. Let O_θ be the point of intersection of $A_\theta D_\theta$ and $B_\theta E_\theta$. Note that O_θ necessarily lies on $C_\theta F_\theta$. Define $f(t) = O_\theta C_\theta - O_\theta B_\theta$. Since $A_{180°} = D_{0°}$, $B_{180°} = E_{0°}$, $C_{180°} = F_{0°}$, $D_{180°} = A_{0°}$, $E_{180°} = B_{0°}$ and $F_{180°} = C_{0°}$, $f(180°) = -f(0°)$. If $f(0°) = 0$, then $A_{0°}B_{0°}C_{0°}D_{0°}E_{0°}F_{0°}$ is a hexagon with the desired property. If not, we have $f(0°) < 0 < f(180°)$ or $f(0°) > 0 > f(180°)$. In either case, $f(\theta) = 0$ for some *theta* in $(0°, 180°)$, and $A_\theta B_\theta C_\theta D_\theta E_\theta F_\theta$ is a hexagon with the desired property.

Kovner's Theorem.

In each strictly convex figure $\mathcal{F}$, one can construct a centrally symmetric figure whose area is at most $\frac{2}{3}$ that of $\mathcal{F}$, and this is best possible.

Proof:

By Theorem 19, a hexagon $ABCDEF$ can be inscribed in the figure where AB, DE and CF are parallel, BC, AD and EF are parallel, CD, BE and FA are parallel, and AD, BE and CF are concurrent. Note that $ABCDEF$ is centrally symmetric. Now AD, BE and CF partition the hexgaon into six congruent triangles, each of area say 1. Then the area of the hexagon is 6, and we shall show that the area of the convex figure is at most 9. The extensions of the sides of the hexagon define six other triangles of area 1. Since these twelve triangles cover the convex figure, its area is at most 12. Let the extensions of CB and FA meet at P, and the extensions of BA and EF meet at Q. Since A is a boundary point of the convex figure, there is a supporting line through A by Theorem 11. Let it cut PB at R and QF at S. Now triangles APR and AQS are not needed for covering the convex figure. Since triangles APR and AFS are congruent, their total area is 1. Hence the area of the convex figure is at most 11. Similar reductions around the vertices C and E yield the desired result. To see that the result is best possible, consider an equilateral triangle ABC of side 1. Suppose a centrally symmetric figure is inside ABC. Let O be the center of symmetry, and let DEF be the image of ABC under a half-turn about O. Then the figure is also inside DEF. To maximize its area, we may as well take it to be the intersection of ABC and DEF. It is easy to see that this intersection is not the largest unless it is a hexagon. This is obtained by removing three equilateral triangles from ABC. Let x, y and z be their respective side lengths. Then $x+y+z=1$. The total area of these triangles is proportional to $x^2 + y^2 + z^2$. By the Power Means Inequality, we have $\frac{x^2+y^2+z^2}{3} \geq \frac{(x+y+z)^2}{9} = \frac{1}{9}$ or $x^2 + y^2 + z^2 \geq \frac{1}{3}$, with equality if and only if $x = y = z$. It follows that the area of the hexagon is at most $\frac{2}{3}$ the area of ABC.

The largest circle that can fit within a convex figure is said to be the **incircle** of the figure. The center and the radius of the incircle are called the **incenter** and the **inradius** respectively.

Note that the definition of incircle here is still different from the definition of the Euclidean incircle of a polygon. Not every polygon has an Euclidean incircle, but every figure has an incircle. Also, whenever a polygon has an Euclidean incircle, it is always unique. This is not the case with convex figures. For example, a non-square rectangle has infinitely many incircles.

Theorem 20.
If the incircle of a convex figure contains a boundary point of the figure, then the tangent to the incircle at this point is a supporting line of the figure.

Proof:
Let P be a boundary point of a convex figure which lies on its incircle. Let ℓ be a supporting line of the figure through P. Then ℓ must also be a supporting line of the incircle, which means that ℓ is in fact a tangent to the incircle.

Theorem 21.
The incircle of a convex figure contains either two boundary points of the figure that are diametrically opposite or three boundary points of the figure that form an acute triangle. In the latter case, the incircle is unique.

Proof:
If the incircle is bounded by two parallel supporting lines, then the points of tangency are diametrically opposite. If it is bounded by at least three supporting lines, not all of the points of tangency can be in the same semicircle. Hence three of them form an acute triangle DEF. Then the tangents to the incircle at D, E and F form a triangle ABC. By Theorem 20, BC, CA and AB are supporting lines of the figure, so that the figure lies within ABC. It is easily seen that the incircle of the figure is also the Euclidean incircle of the triangle, which is the unique largest circle within ABC.

Blaschke's Theorem.
If r is the inradius of a convex figure of breadth 1, then $\frac{1}{3} \le r \le \frac{1}{2}$.

Proof:
Consider any incircle of the figure. Its diameter is at most the breadth of the figure, so that $r \le \frac{l}{2}$. This is attained when the figure is a circle of diameter 1. We now prove the lower bound. Consider any incircle $\mathcal{K}$. Suppose it contains two boundary points P and Q of the figure which are diametrically opposite. By Theorem 20, the tangents to $\mathcal{K}$ at P and Q are supporting lines. They are parallel to each other and at a distance $2r$ apart. Hence $2r \ge 1$ so that $r \ge \frac{1}{2} > \frac{l}{3}$. Otherwise, by Theorem 21, $\mathcal{K}$ must contain three boundary points of the figure that form an acute triangle DEF. By Theorem 20, the tangents to $\mathcal{K}$ at D, E and F form a triangle ABC which encloses the figure. Let $a = BC$, $b = CA$ and $c = AB$. Let h_a, h_b and h_c be the respective altitudes of triangle ABC. Then twice the area of ABC is given by $(a+b+c)r = ah_a = bh_b = ch_c$. By symmetry, we may assume that $h_a \le h_b \le h_c$. We must have $h_a \ge l$. Now $3(a+b+c)r = ah_a + bh_b + ch_c \ge (a+b+c)h_a$ so that $r \ge \frac{h_a}{3} \ge \frac{1}{3}$. This is attained when the figure is an equilateral triangle of side length $\frac{2}{\sqrt{3}}$.

Exercises

4. Two bugs are crawling at equal constant speed in the clockwise direction once round the boundary of a convex figure. What should their starting positions be if the minimum distance between them during the crawl is to be as large as possible?

5. Prove that a figure $\mathcal{F}$ is convex if and only if for any exterior point P of $\mathcal{F}$, there is a unique point of $\mathcal{F}$ which is closest to P.

6. Prove that a square can be inscribed in any strictly convex figure.

Section 3. Figures of Constant Width

Why is a man-hole cover round? The simple answer is that the man-hole is round, but that begs the question: Why is the man-hole round? Often, it is said that the man is round, but a round man can get down a square hole just as easily. It is then said that it is impossible to drill a square hole.

As it turns out, it is possible to drill a square hole. The square man-hole is undesirable for another reason. If we hold up the square man-hole cover so that its edge is lined up with the diagonal of the man-hole, the cover can be dropped into the man-hole although its size is larger than that of the man-hole. This has severe consequences for a man working below the man-hole.

A round man-hole cover which is slightly larger cannot be dropped into a round man-hole. This is because the width of the circle is the same in every direction. Such a figure is called a **figure of constant width**.

Theorem 22.
A figure of constant width has central symmetry if and only if it is a circle.

Proof:
Only the converse requires an argument. Suppose a figure of constant width 1 has a center of symmetry O. Consider any pair of parallel supporting lines. They are at a distance 1 apart, which is a maximum. By Theorem 8, each contains exactly one point of the figure, say A and B respectively, such that AB is perpendicular to the supporting lines. We claim that O lies on AB. If not, let A' and B' be the respective images of A and B under central symmetry about O. Then $A'B' = AB = 1$, and $A'B'$ is parallel to AB. This contradicts Theorem 13. Hence O lies on AB. Since O is the center of symmetry, it is the midpoint of AB. Hence the two supporting lines are at a distance $\frac{1}{2}$ from O. Since they are arbitrarily chosen, every supporting line of the figure is at a distance $\frac{1}{2}$ from O. It follows easily that the figure is a circle with center O and radius $\frac{1}{2}$.

Theorem 23.
A figure of constant width is strictly convex.

Proof:
Suppose a support line contains two boundary points A and B of a figure of constant width. Let C be any point on the support line parallel to AB. Then one of AB and AC will exceed in length the width of the figure, which is a contradiction. Hence each support line contains exactly one boundary point of the figure. In other words, the figure is strictly convex.

Theorem 24.
A boundary point is a corner point of a figure of constant width if and only if it is the intersection of two diameters.

Proof:

Suppose AB and AC are diameters. Then the lines through A perpendicular respectively to AB and AC are distinct supporting lines by Theorem 8. Hence A is a corner point. Conversely, suppose A is a corner point of a figure of constant width. Then there are two distinct supporting lines through A. Consider the chords AB and AC perpendicular to them respectively. By Theorem 8, the line through B perpendicular to AB must also be a supporting line. Hence AB is a diameter of the figure. Similarly, so is AC.

Theorem 25.

AB and AC are diameters of a figure of constant width 1 if and only if the part of the boundary between B and C and not containing A is an arc with center A and radius 1.

Proof:

Suppose AB and AC are diameters. Clearly, the lines through A perpendicular to AB and AC are supporting lines. For any point D on the part of the boundary in question, the line through A perpendicular to AD is a supporting line of the figure, as is the line through D perpendicular to AD. Since the figure is of constant width 1, $AD = 1$. Since D is arbitrarily chosen, the part of the boundary in question must be an arc with center A and radius 1. We now prove the converse. We first show that A is a boundary point of the figure. Clearly, the line through B perpendicular to AB is a supporting line of the figure. Consider the other supporting line parallel to it. Since the figure is of constant width 1, the distance between these two supporting lines is 1. Since $AB = 1$, the other supporting line passes through A. By Theorem 8, A is a boundary point and AB is a diameter of the figure. Similarly, AC is also a diameter.

Theorem 26.

The interior angle at a corner point of a figure of constant width cannot be less than $120°$.

Proof:

Suppose A is a corner point of a figure of constant width. Let b and c be supporting lines through A which define the interior angle at A. Let AB and AC be the chords perpendicular to b and c respectively. Then AB and AC are diameters. Hence $\angle BAC \le 60°$ as otherwise $BC > AB$. Now the exterior angle at A is equal to $\angle BAC$. Hence the interior angle at A cannot be less than $l20°$.

We have gone through a list of properties of figures of constant width. Still, apart from the circle, it is not easy to see that there exist other examples. We now introduce a prototype.

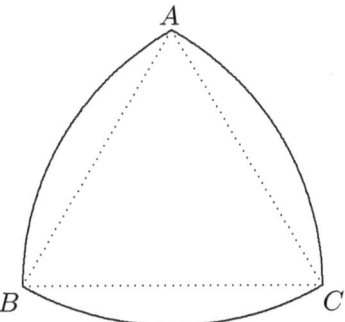

Figure 5.7

Let ABC be an equilateral triangle. Draw minor arcs with centers A, B and C joining B to C, C to A and A to B respectively. The figure enclosed by these three arcs, called a **Reuleaux triangle**, is shown in Figure 5.7.

Theorem 27.
A Reuleaux triangle is a figure of constant width.

Proof:
We may assume that the Reuleaux triangle is constructed from an equilateral triangle ABC of side length 1. Then its diameter is 1. Consider a pair of parallel support lines. Since it is strictly convex, each line contains exactly one boundary point of the figure. If both points are among the corner points A, B and C, then the distance between them is 1. Hence the width in this direction is at least 1. If one of the points is a regular boundary point, say on the arc BC. Then the support line at this point is the tangent to the arc BC, which has center A. Again the width in this direction is at least 1. However, no width can exceed the diameter, so that the width is exactly 1, and this is the constant width in every direction.

Theorem 28.
The only figure of constant width with a corner point having an interior angle of 120° is the Reuleaux triangle.

Proof:
Suppose a figure of constant width has a corner point A with an interior angle of 120°. Let b and c be supporting lines through A which define this angle. The chords AB and AC, perpendicular to b and c respectively, are diameters. Moreover, $\angle BAC = 60°$ since it is equal to the exterior angle. It follows that BC is also a diameter. The result now follows from Theorem 25.

Beginning with Reuleaux triangles, it is easy to construct other figures of constant width. Extend the sides of an equilateral triangle ABC by the same length, BA to A_b, CA to A_c, CB to B_c, AB to B_a, AC to C_a and BC to C_b. The **modified** Reuleaux triangle is obtained by drawing six circular arcs, as shown in Figure 5.8. We can prove as in Theorem 27 that this is a figure of constant width B_cC_b.

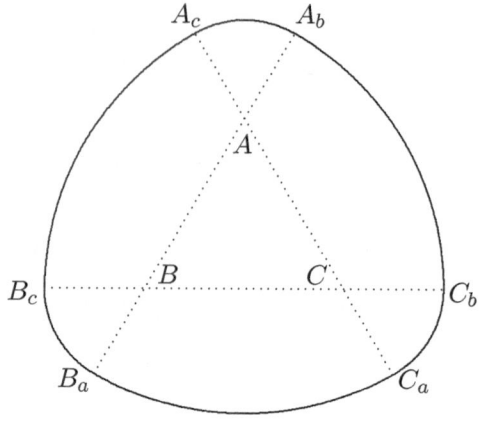

Figure 5.8

Similarly, **Reuleaux $(2n+1)$-gons** and modified Reuleaux $(2n+1)$-gons can be constructed from a regular $(2n + 1)$-gon. We may in fact replace the regular $(2n + 1)$-gon by a $(2n + 1)$-gon $A_1A_2 \ldots A_{2n+1}$ which satisfies the condition $A_1A_{n+1} = A_2A_{n+2} = \cdots = A_{2n+1}A_n$. The resulting figure is called a **generalized** Reuleaux $(2n + 1)$-gon. We can also have a modified generalized Reuleaux $(2n+1)$-gon. It is easy to see that all these are figures of constant width.

We now deal with results on circles associated with figures of constant width.

Theorem 29.
If a circle has at least three points in common with a figure of constant width 1, then the radius of the circle is at most 1.

Proof:
Suppose a circle γ has three points P, Q and R in common with a figure of constant width 1. We may assume $\angle QPR \leq \angle PQR$ and $\angle QPR \leq \angle PRQ$. Through P, draw a supporting line ℓ to the figure and draw the circle of radius 1 tangent to ℓ at P. By Theorem 13, this circle contains Q and R. Extend PQ and PR to meet this circle at Q' and R' respectively. We claim that $Q'R' \leq QR$, from which it follows that the radius of y is at most 1. If $Q = Q'$ and $R = R'$, there is nothing more to prove. Suppose $Q = Q'$ but $R \neq R'$. Then $\angle QRR'$ is obtuse so that $Q'R' > Q'R = QR$. The case $Q \neq Q'$ and $R = R'$ is handled in the same way.

Finally, suppose $Q \neq Q'$ and $R \neq R'$. Since $\angle PQR + \angle PRQ = \angle PQ'R' + \angle PR'Q'$, we may assume that $\angle PQR \leq \angle PQ'R'$. Now $Q'QR' > \angle QPR \geq \angle PQR \geq \angle PQ'R'$ so that $Q'R' > QR'$. Since $\angle QRR'$ is obtuse, $QR' > QR$.

Theorem 30.
The circumcircle and the incircle of a figure of constant width 1 are concentric and the sum of the circumradius and the inradius is 1.

Proof:
Let O be the center of any circle γ enclosed in the figure $\mathcal{F}$, with radius r. We claim that the circle with center O and radius $1 - r$ encloses $\mathcal{F}$. Otherwise, there is a point P of $\mathcal{F}$ outside this circle. Extend PO to meet γ at Q. Now $PQ = OP + OQ > (1 - r) + r = 1$. This is a contradiction. Similarly, if a circle of radius R encloses $\mathcal{F}$, then the concentric circle of radius $1 - R$ is enclosed in $\mathcal{F}$. Now let R and r be the circumradius and the inradius of $\mathcal{F}$ respectively. Then $R \leq 1 - r$ since the circle of radius $l - r$ concentric with the incircle encloses $\mathcal{F}$, and the circumcircle is the smallest circle enclosing $\mathcal{F}$. Similarly, $r \geq 1 - R$. Hence $R + r = 1$. Finally, if these circles are not concentric, then the circle of radius R concentric with the incircle will also be a circumcircle. This contradicts the uniqueness of the circumcircle.

Corollary.
The incircle of a figure of constant width is unique.

Theorem 31.
Among all figures of constant width 1, the circle has the smallest circumradius and the Reuleaux has the greatest circumradius.

Proof:
A figure of constant width 1 has diameter 1. By Jung's Theorem, its circumradius lies between $\frac{1}{2}$ and $\frac{1}{\sqrt{3}}$. The circumradius of the circle is clearly $\frac{1}{2}$, and it is easy to verify that the circumradius of the Reuleaux triangle is $\frac{1}{\sqrt{3}}$. If the figure is not a circle, then its circumradius is clearly greater than $1/2$. Suppose the figure has circumradius $\frac{1}{\sqrt{3}}$. Then the circumcircle must contain three points of the figure which form an acute triangle. If it is not an equilateral triangle of side 1, it is easy to see that its circumradius is strictly less than $\frac{1}{\sqrt{3}}$. If it is an equilateral triangle of side 1, then the figure must be the Reuleaux triangle.

Corollary.
Among all figures of constant width 1, the circle has the greatest inradius and the Reuleaux has the smallest inradius.

We conclude with some results on the perimeters and areas of figures of constant width.

Theorem 32.
Let $ABCD$ be a rhombus with $BD = a$, where $1 \le a \le 2$. Let MN and PQ be chords perpendicular to BD at a distance 1 apart, with MN closer to B.

(a) $MN + PQ - BM - BN - DP - DQ$ does not depend on the positions of MN and PQ.

(b) The total area of triangles BMN and DPQ is minimum when MN and PQ are equidistant from AC, and maximum when one of them passes through B or D. In all other positions of MN and PQ, the total area lies strictly between the maximum and minimum values.

Proof:
Let h denote the distance from B to MN and k denote the distance from D to PQ. Then we have $h + k = a - 1$. Let α denote $\angle ABD$.

(a) We have $BM = h \sec \alpha$, $DP = k \sec \alpha$, $MN = 2h \tan \alpha$ and $PQ = 2k \tan \alpha$. Hence $MN + PQ - BM - BN - DP - DQ = 2(\tan \alpha - \sec \alpha)(a - 1)$, which is constant.

(b) The areas of triangles BMN and DPQ are given by $h^2 \tan \alpha$ and $k^2 \tan \alpha$ respectively, and the total is $\frac{\tan \alpha}{2}((a - 1)^2 + (h - k)^2)$. The minimum occurs if and only if $h = k$, that is, when MN and PQ are equidistant from AC. The maximum occurs if and only if $h = 0$ or $k = 0$, that is, when MN passes through B or PQ passes through D.

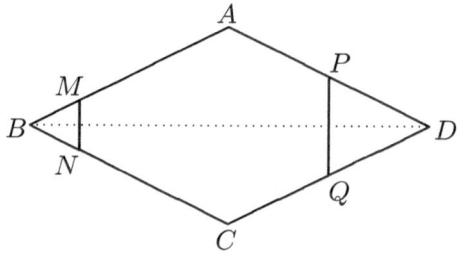

Figure 5.9

Barbier's Theorem.
All figures of constant width 1 has perimeter π.

Proof:
Consider the circle and any other figure of constant width 1. By Theorem 5, a square of side 1 can be circumscribed about each. We now reduce each square to a circumscribing equiangular 2^n-gon by cutting off pairs of opposite vertices. Let B and D be the two opposite vertices to be cut off at a given stage. Note that $1 < \frac{BD}{\sqrt{2}} < 2$. Extend the sides at B and D, if necessary, to form a parallelogram $ABCD$, which is a rhombus since the two pairs of opposite sides are at equal distance apart. We draw supporting lines of the figure perpendicular to BD, forming the chords MN and PQ of the rhombus, with MN closer to B. We now cut off triangles BMN and DPQ. The net change in perimeter is $MN + PQ - BM - BN - DP - DQ$, which is constant by Theorem 32. On passage to limit, both figures will have the same perimeter, and that of the circle is clearly π.

Blaschke-Lebesque Theorem.
Among all figures of constant width 1, the circle has the greatest area and the Reuleaux triangle has the smallest area. All others have areas lying strictly between the maximum and minimum values.

Proof:
Consider the circle, the Reuleaux triangle and any other figure of constant width 1. By Pal's Second Theorem, a regular hexagon of side $\frac{1}{\sqrt{3}}$ can be circumscribed about each. Note that three of the vertices of the hexagon circumscribed about the Reuleaux triangle coincide with the corner points of the Reuleaux triangle. We now reduce each hexagon to a circumscribing equiangular dodecagon by cutting off pairs of opposite vertices. Let B and D be two opposite vertices to be out off. Note that $1 < BD \leq \frac{2}{\sqrt{3}} < 2$. Extend the sides at B and D to form a parallelogram $ABCD$. Constant width implies that this is a rhombus. We draw supporting lines of the figure perpendicular to BD, forming the chords MN and PQ of the rhombus, with MN closer to B. We now cut off triangles BMN and DPQ. When the figure is the circle, MN and PQ will be equidistant from AC. When the figure is the Reuleaux triangle, either MN passes through B or PQ passes through D. Moreover, the dodecagon circumscribing the Reuleaux triangle is degenerate. It has only nine sides of positive lengths and six of its twelve vertices are at the corner points of the Reuleaux triangle. We continue to cut off opposite vertices so that each figure is approximated by a sequence of equiangular $3 \cdot 2^n$-gons. At each stage, the area removed from the polygon circumscribing the circle is the smallest while the area removed from the polygon circumscribing the Reuleaux triangle is the greatest. On passage to limit, the circle has the greatest area and the Reuleaux triangle has the smallest area.

By Theorem 22, the circle is the only figure of constant width with central symmetry. For any other figure of constant width, the area removed from the polygon circumscribing it is at least as great as that removed from the polygon circumscribing the circle, and strictly greater on at least one occasion. Hence its area is smaller than that of the circle. For any figure of constant width, the area removed from the hexagon circumscribing it is at most as great as that removed from the hexagon circumscribing the Reuleaux triangle, and equality can only hold if one vertex of the hexagon is a corner point of the figure. The interior angle of the figure at this point is at most 120°, so that it is exactly 120° by Theorem 26. By Theorem 28, this figure can only be the Reuleaux triangle itself.

Returning to our opening remark about man-hole covers, the Reuleaxu triangle is the key to drilling a square hole. This is because it can rotate inside a square whose side length is equal to its constant width. By removing parts of it without losing contact with the sides of the square, we can make sharp cutting corners in it. An axis is attached to the center of the Reuleaux triangle, but it is not fixed in place as otherwise we will be drilling a round hole once again. Instead, the axis is constrained to rock about within a cylinder. This device is called a Watt's drill.

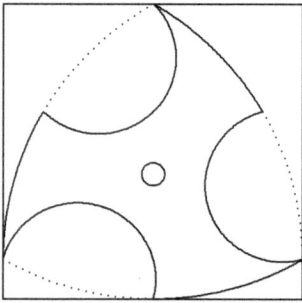

Figure 5.10

Exercises

7. Prove that if a figure of constant width 1 whose boundary consists of a union of circular arcs of radius 1 has n corner points, where n is a positive integer, then n is odd.

8. A convex figure $\mathcal{F}$ is bounded by a line segment AB of length 1 and a curve γ joining A and B and lying between the lines perpendicular to AB through A and B respectively. Moreover, for each point P on γ and each supporting line ℓ at P, there is a circle of radius 1 tangent to γ at P which encloses $\mathcal{F}$. Prove that there is a unique curve joining A and B such that it and γ enclose a figure of constant width 1.

9. Construct a figure of constant width with exactly two corner points.

Bibliography

1. H. Hadwiger, H. Debrunner and V. Klee, Combinatorial Geometry in the Plane, Holt, Rinehart & Winston, New York (1964).

2. I. M. Yaglom and W. G. Boltyanski, Convex Figures, Holt, Rinehart & Winston, New York (1956).

Solution to Exercises

1. By Theorem 1, for any direction, there is a line which bisects its area. Let the initial direction be horizontal and pointing to the right. Then it makes a $0°$ angle with the positive x-axis. If the perimeter is also bisected, there is nothing further to prove. If not, rotate the line continuously (but not necessarily about a fixed point) so that it makes an angle θ with the positive x-axis and bisecting the area of the figure. Define $f(\theta)$ to be the difference when the total length of the part of the perimeter of the figure on the left side of the line is subtracted from the total length of the part of the perimeter of the figure on the right side of the line. Since $f(0°) \neq 0$, $f(180°) \neq 0$ either since it is the same line. On the other hand, the direction is now pointing to the left, so that one of $f(180°)$ is opposite in sign to $f(0°)$. It follows from the Intermediate Value Theorem that $f(\theta) = 0$ for some θ between $0°$ and $180°$. The desired conclusion follows.

2. Suppose an equiangular pentagon has an axis of symmetry. Since it has an odd number of vertices, exactly one of them must lie on the axis. Then the two sides which meet there have equal lengths. Conversely, if two adjacent sides of an equiangular pentagon have equal lengths, then the pentagon has an axis of symmetry passing through their common vertex. By Theorem 4, there exists a unique equiangular pentagon $A_\theta B_\theta C_\theta D_\theta E_\theta$, going counterclockwise, circumscribed about the figure and with $D_\theta E_\theta$ making an angle θ with the positive x-axis. Define $f(\theta) = A_\theta B_\theta - A_\theta E_\theta$. If $f(0°) = 0$, then $A_{0°} B_{0°} C_{0°} D_{0°} E_{0°}$ is a pentagon with the desired property. If not, we may assume that $f(0°) > 0$ so that $A_0 B_0 > A_0 E_0$. If $f(72°) \leq 0$, we have the desired pentagon for some θ in $(0°, 72°]$. Otherwise, $B_0 C_0 > A_0 B_0$, but we cannot have $A_0 E_0 < A_0 B_0 < B_0 C_0 < C_0 D_0 < D_0 E_0 < E_0 A_0$. Hence one of $f(144°)$, $f(216°)$ and $f(288°)$ must be non-positive. It follows that $f(\theta) = 0$ for some θ in $(72°, 288°]$, and $A_\theta B_\theta C_\theta D_\theta E_\theta$ is a pentagon with the desired properties.

3. That this is best possible can be seen if we take the figure to be a circle of radius $\frac{1}{2}$. By Theorem 5, there exists a square circumscribed

about the figure. This square is of side at most 1 since the figure has diameter 1. Clearly a square of side 1 can cover the figure.

4. We choose the starting points A_0 and B_0 such that the distance between them along the boundary is exactly half the perimeter of the convex figure. Then their positions A_t and B_t at any time t also have the same property, and could have been chosen as the starting points. Suppose C_0 and D_0 are starting points such that the distance between them along the boundary is not equal to half the perimeter. Then this also holds true for their positions C_t and D_t at any time t. Let w be the time when the length $A_w B_w$ is minimum. Then there are times u and v such that $C_u D_u$ and $C_v D_v$ are both parallel to $A_w B_w$, one of each side of it. Since $A_w B_w$ cannot be shorter than both $C_u D_u$ and $C_v D_v$, $A_w B_w$ is not shorter than the minimum length of $C_t D_t$.

5. Suppose $\mathcal{F}$ is convex. Let P be any exterior point of $\mathcal{F}$. Suppose there are two points Q and R of $\mathcal{F}$ such that $PQ = PR$ is the minimum distance from P to $\mathcal{F}$. Let T be the midpoint of QR. Then T belongs to $\mathcal{F}$ since $\mathcal{F}$ is convex. However, $PT < PQ$, and we have a contradiction. Conversely, suppose that for any point P of $\mathcal{F}$, there is a unique point of $\mathcal{F}$ which is closest to P. If $\mathcal{F}$ is not convex, then there exist two two boundary points Q and R of $\mathcal{F}$ such that QR is a supporting line of $\mathcal{F}$ but every point between Q and R is an exterior point of $\mathcal{F}$. Let $\mathcal{F}$ lie below QR. Let U be the boundary point of $\mathcal{F}$ between Q and R that lies furthest below QR, at a depth of h. Let $QR = d$. Construct the rectangle $QRST$ above QR, such that $QT = \frac{d^2}{2h}$. Let P be a point moving from T to S. When $P = T$, clearly Q is the point in $\mathcal{F}$ closest to P. Similarly, it is R when $P = S$. As P moves continuously from T to S, the point in $\mathcal{F}$ closest to P must move continuously from Q to R along the boundary of $\mathcal{F}$. At some point in time, U will be the point of $\mathcal{F}$ closest to P which is somewhere between T and S. However, $PU > \frac{d^2}{2h} + h = \sqrt{(\frac{d^2}{2h})^2 + d^2 + h^2} > \sqrt{(\frac{d^2}{2h})^2 + d^2} = SQ > PQ$, so that Q is closer to P then U. This is a contradiction.

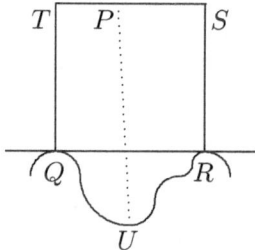

Figure 5.11

6. Let $C_\theta^d D_\theta^d$ be a chord of the figure, with length d and making an angle θ with the positive x-axis. Let $A_\theta^d B_\theta^d$ be a chord of equal length but in the opposite direction to $C_\theta^d D_\theta^d$. Let E_θ^d be the point of intersection of the diagonals of the parallelogram $A_\theta^d B_\theta^d C_\theta^d D_\theta^d$. Keeping θ fixed for now, let d vary from 0 to the maximum length m of a chord in the same direction. Define $f_\theta(d) = \angle A_\theta^d E_\theta^d B_\theta^d - \angle B_\theta^d E_\theta^d C_\theta^d$. Note that we have $f_\theta(0) = -180°$ while $f_\theta(m) = 180°$. Hence $f_\theta(d) = 0°$ for some d in $(0, m)$. In this position, the parallelogram $A_\theta B_\theta C_\theta D_\theta$ is a rhombus, and it is unique for the given direction. We now define $g(\theta) = \angle A_\theta B_\theta C_\theta - \angle B_\theta C_\theta D_\theta$. Note that if $g(0) = 0°$, then $A_0 B_0 C_0 D_0$ is the desired square. Suppose $g(0) < 0°$ and $A_\theta D_\theta$ makes an angle ϕ with the positive x-axis. Then $A_\phi = B_{0°}$, $B_\phi = C_{0°}$, $C_\phi = D0°$ and $D_\phi = A_{0°}$. Hence $g(\phi) > 0°$ and we must have $g(\theta) = 0°$ for some θ in $(0°, \phi)$. It follows that for that value of θ, $A_\theta B_\theta C_\theta D_\theta$ is the desired square.

7. Suppose to the contrary that a figure of constant width 1 has a non-zero even number of corner points A_1, A_2, $\ldots$, A_n. We claim that for each A_i, $A_i A_j$ is a diameter for exactly two values of j. Suppose $A_i A_j$, $A_i A_k$ and $A_i A_\ell$ are all diameters, say with $A_i A_k$ between the other two. By Theorem 25, A_k lies on a circular arc with center A_i and radius 1 joining A_j and A_ℓ. This is impossible since A_k will not be a corner point. On the other hand, by Theorem 24, A_i is the intersection of two diameters $A_i B$ and $A_i C$. By Theorem 25, the part of the boundary between B and C is a circular arc with center A_i and radius 1. We may choose B and C so that they are the endpoints of the circular arc. Then each of B and C is a corner point, and is equal to A_j for some j. Moreover, note that if $A_i A_j$ and $A_i A_k$ are diameters, then A_j and A_k must be consecutive. Suppose $A_1 A_j$ and $A_1 A_{j+1}$ are diameters. There are $j - 2$ corner points between A_1 and A_j, and $n - j - 1$ corner points between A_{j+1} and A_1. Since n is even, these two numbers cannot be equal. We may assume that $j - 2 > n - j - l$. Now each A_i, $2 \le i \le j - 1$, is the intersection of two diameters. By Theorem 13, both of these must terminate at A_k for some k, $j + l \le k \le n$. Note that $A_1 A_{j+1}$ is a diameter. Since $2(j - 2) > 2(n - j - 1) + l$, we have a contradiction.

8. With A and B as respective centers, draw two minor arcs of radius 1 intersecting at C and D. Let $\mathcal{F}$ be on the same side of AB as C. By hypothesis, $\mathcal{F}$ is a subset of the figure $\mathcal{G}$ enclosed by AB and the arcs AC and BC. Let $\mathcal{G}'$ denote the figure enclosed by AB and the arcs AD and BD. Note that $\mathcal{F} \cup \mathcal{G}'$ is a convex figure. Consider all circles of radius 1 with centers on γ. The intersection $\mathcal{H}$ of $\mathcal{F} \cup \mathcal{G}'$ and this collection of circles is convex. We now show that H is a figure of constant width 1 having γ as part of its boundary.

Let P be any point on γ. Since $\mathcal{F}$ is a subset of $\mathcal{G}$, P is at a distance at most 1 from any other point on γ. Hence P belongs to every circle of radius 1 with center on γ, so that P is in $\mathcal{H}$. Since P is a boundary point of $\mathcal{F}$, it must be a boundary point of $\mathcal{H}$. Hence $\mathcal{H}$ has γ as part of its boundary. Since $\mathcal{H}$ is convex, $\mathcal{F}$ is a subset of $\mathcal{H}$. No two points of $\mathcal{H}$ can be further apart than the distance 1. Note that $\mathcal{H}$ is a subset of $\mathcal{F} \cup \mathcal{G}'$. If two points of $\mathcal{H}$ are both in $\mathcal{F}$ or both in $\mathcal{G}'$, there is nothing more to prove. Suppose P is in $\mathcal{F}$ and P' is in $\mathcal{G}'$. We may assume that P is on γ. Since P' is in $\mathcal{H}$ and $\mathcal{H}$ is contained in the circle of radius 1 with P as center, we have $PP' \leq 1$. Hence the diameter of $\mathcal{H}$ is 1. We now show that its breadth is also 1. Consider any pair of parallel supporting lines of $\mathcal{H}$ not perpendicular to AB. One of them will have a point P of contact with γ. Let PQ be perpendicular to the supporting lines with $PQ = 1$. The proof will be completed if we can show that Q is in $\mathcal{H}$. By hypothesis, the circle with Q as center and having radius 1 encloses $\mathcal{F}$. Hence the distance between Q and any point on γ is at most 1, so that Q belongs to every circle of radius 1 with center on γ. In particular, $AQ \leq 1$ and $BQ \leq 1$. Since Q lies on the opposite side of AB to P, Q lies in $\mathcal{G}'$ and therefore in $\mathcal{F} \cup \mathcal{G}'$. Hence Q is indeed in $\mathcal{H}$.

9. Take $AB = AC = BD = 1$. Let AC intersect BD at E such that $AE = BE = 0.625$. With A, E, B and E again as centers, draw minor arcs joining B to C, C to D, D to A and A to B, respectively. This figure has constant width 1, with only A and B as corner points. C is not a corner point since it is collinear with the centers of both arcs which meet there. Similarly, D is not a corner point either.

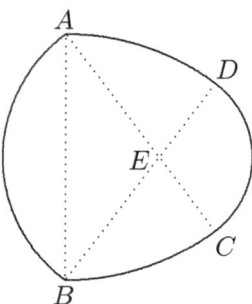

Figure 5.12

Part II: Other Topics

Chapter Six: Balancing Problems

Section 1. Identifying Fake Coins

In the standard problem, several coins are given. Most of them are real, and all real coins have the same weight. The weight of a fake coin is different from that of a real coin. It may be heavier or lighter. If there are two or more fake coins, they do not necessarily have the same weight. Our task is to identify the fake coins, and determine whether they are heavier or lighter than real coins if that is not given.

Our sole instrument is a standard balance. It consists of two pans of negligible weight, attached by strings of equal length and negligible weight, to a straight beam pivoted at a point halfway between the points of suspension of the pans. Various objects can be placed on either pan, and the total weight of the objects in a pan is called the weight of that pan. If the two pans have equal weights, the balance is said to be in equilibrium and the beam is horizontal. Otherwise, the lighter pan will rise higher than the heavier pan.

Our primary source of puzzles is the International Mathematics Tournament of the Towns. See [3] to [8]. See also [1]. However, we start with a classic puzzle.

Problem 1.
We have nine coins, one of which is fake. The fake coin weighs more than a real coin. Identify the fake coin in two weighings.

Solution:
Let the coins be A, B, C, D, E, F, G, H and I. In the first weighing, put ABC in the first pan and GHI in the second pan. If we have equilibrium, the fake coin is among D, E and F. In the second weighing, we put D in the first pan and F in the second pan, and we can identify the fake coin. If we do not have equilibrium in the first weighing, the fake coin is among the three coins in the heavier pan. We can identify it in the second weighing as before.

The solution to the first problem is called an *adaptive* solution. What we do in the second weighing depends on the outcome of the first weighing. In a *non-adaptive*, what we do in the second weighing is determined in advance.

The following puzzle appears without solutions in many books. See for instance [2]. We offer a non-adaptive solution here.

© Springer International Publishing AG 2018
A. Liu, *S.M.A.R.T. Circle Minicourses*, Springer Texts in
Education, https://doi.org/10.1007/978-3-319-71743-2_6

Problem 2.

There are twelve coins, one of which is fake. It is not known whether the fake coin is heavier or lighter than a real coin. Identify the fake coin and determine whether it is heavier or lighter in three weighings.

Solution:

Label the coins A, B, C, D, E, F, G, H, I, J, K and L. Our plan is to weigh four of the coins against four others in each weighing. To do so systematically, we assign codes to each coin. In a weighing, there are three possible outcomes: either pan can go up, or there may be equilibrium. For this reason, we use numbers in base 3 for our codes. Since we have three weighings, we look at three-digit numbers, of which there are 27. In base 3, if we fill in the leading 0s, they are 000, 111, 222 and the 12 pairs in the following chart.

Coin	A	B	C	D	E	F
Primary Code	001	010	011	012	112	120
Secondary Code	221	212	211	210	110	102
Coins	G	H	I	J	K	L
Primary Code	121	122	200	201	202	220
Secondary Code	101	100	022	021	020	002

The fake coin is one of the twelve, and it may be heavy or light. Thus there are twenty-four possible outcomes in our investigation. With twenty-seven available numbers, it makes sense to discard the three with identical digits, and assign two of the others to each coin. The Primary Code for the coins have the following property. There are four coins with 0 as the first digits, four with 1 as their first digits, and four with 2 as their first digits. The same applies to their second and third digits. The Secondary Code for the coins are the *complements* of their Primary Codes, in that the 1s remain the same while the 0s and the 2s are interchanged. How is this chart constructed? There are no hard and fast rules. The scheme we have adopted is as follows. Since we have discarded the three numbers with identical digits, each of the remaining 24 numbers contains two different digits. Check the change over for the first time, which may occur between the first and the second digits, or between the second and the third digits. If it is a change from 0 over to 1, 1 over to 2 or 2 over to 0, we use the number as a Primary Code. If it is a change from 0 over to 2, 2 over to 1 or 1 over to 0, we use the number as a Secondary Code.

For now, assume that the fake coin is heavy. In the first weighing, place the four coins with 0 as the first digits of their Primary Codes in one pan and those with 2 as the first digits of their Primary Codes in other pan. In the second weighing, we look at the second digits of their Primary Codes, and in the third weighing, we look at the third digits of their Primary Codes. In both cases, as with the first weighing, we put the four 0s up against the four 2s.

Thus we can determine the Primary Code of the fake coin, on the assumption that it is heavy. If this is indeed the Primary Code of a coin, we have identified the fake coin, and it is heavy. If instead this is the Secondary Code of a coin, we still have identified the fake coin, but it is light.

We can extract an adaptive solution from the above. In the first weighing, we weigh A, B, C and D against I, J, K and L. Suppose there is equilibrium. In the second weighing, we weigh I, J and K against F, G and H. Suppose there is equilibrium once again. Then the fake coin is E. In the third weighing, we weigh L against E to see if E is light or heavy. Suppose in the second weighing, F, G and H are heavier. Then one of them is the fake coin, and it is heavy. In the third weighing, we weigh F against H, and will know everything. Suppose in the first weighing, I, J, K and L are heavier. In the second weighing, we weigh A, I, J and K against F, G, H and L. If there is equilibrium, the fake coin is among B, C and D and it is light. It can be found in the third weighing when we weigh B against D. If A, I, J and K are heavier, the fake coin is among I, J and K and it is heavy. It can be found in the third weighing when we weigh I against K. If F, G, H and L are heavier, either A is light or L is heavy. In the third weighing, we weigh E against L to see if L is real or fake.

We now tackle some miscellaneous puzzles.

Problem 3.
There are five coins, two of which are fake. The fake coins have different weights, and each is heavier than a real coin. Identify both fake coins in three weighings.

Solution:
Let the coins be A, B, C, D and E. Weigh A against B, C against D and D against E. We cannot have equilibrium all three times. If we have it all three times, D must be a fake coin, and whichever of A and B is heavier is the other fake coin. If we have equilibrium twice, they must occur in the second and the third weighings, so that A and B are the fake coins. Suppose we have equilibrium once. If it is in the first weighing, then the fake coins are C and E. Otherwise, we may assume by symmetry that it is in the second weighing. Then E and whichever of A and B is heavier are the fake coins.

Problem 4.
There are six coins, two of which are fake. The fake coins have equal weight, and are lighter than real coins. Identify both fake coins in three weighings.

Solution:
First weigh the coins three against three. If there is equilibrium, we have a fake coin on each side. Now weigh two coins on one side against each other. If there is equilibrium, the remaining coin is fake. Otherwise, the lighter one is fake. The third weighing can determine the pther fake coin.

Suppose the first weighing does not result in equilibrium. Then both fake coins are on the lighter side. Weigh two coins on that side against each other. If there is equilibrium, both are fake. Otherwise, the lighter one and the remaining one are the fake coins.

Problem 5.
We have seven coins, two of which are fake. The fake coins have equal weight, and are lighter than real coins. Identify both fake coins in three weighings.

Solution:
Let the coins be A, B, C, D, E, F and G. In the first weighing, weigh AB against CD. If there is equilibrium, either all four coins are real, or there is a fake coin in each pan. In the second weighing, weigh AC against EF. If there is equilibrium, the fake coins must be B and D. If AC are heavier, A is a fake coin and the other is C or D. It can be identified in the third weighing. If AC are lighter than EF, then both fake coins are among E, F and G. They can be identified in the third weighing. Suppose the first weighing does not result in equilibrium. We may assume by symmetry that AB are heaver. Then at least one of them is fake while both C and D are real. In the second weighing, weigh AC against EF. If there is equilibrium, then the fake coins are either B and G, A and E or A and F. They can be identified in the third weighing where we weigh E against F. If AC are heavier, A is a fake coin and the other one is B or G. It can be identified in the third weighing. Finally, if AC are lighter than EF, A is a real coin while B is fake. The other fake coin is E or F, and can be identified in the third weighing.

Problem 6.
There are five coins, two of which are fake. One of the fake coins is heavier and the other lighter than a real coin. The total weight of the two fake coins is equal to the total weight of two real coins. Identify the heavy fake coin and the light fake coin in three weighings.

Solution:
Let the coins be A, B, C, D and E. In the first two weighings, weigh A against B and C against D. They cannot both result in equilibrium. Suppose there is equilibrium between A and B. Then both are real. We may assume by symmetry that C is heavier than D. This means that either C is heavy and D is light, C is heavy and E is light, or E is heavy and D is light. In the third weighing, weigh C and D against B and E. If there is equilibrium, then C is heavy and D is light. If C and D are heavier, then C is heavy and E is light. If C and D are lighter, then E is heavy and D is light.

Suppose neither of the first two weighings results in equilibrium. We may assume that A is heavier than B and C is heavier than D. This means that either A is heavy and D is light, or C is heavy and B is light. In the third weighing, weigh C and D against B and E. This cannot result in equilibrium. If C and D are heavier, then C is heavy and B is light. If C and D are lighter, then A is heavy and D is light.

Problem 7.

There are sixteen coins, one of which is fake. Identify the fake coin in four weighings.

Solution:

Let the coins be A, B, C, D, E, F, G, H, I, J, K, L, M, N, O and P. In the first weighing, weigh A, B, C, D and E against L, M, N, O and P. If there is equilibrium, weigh F, G and H against I, J and K. We cannot also have equilibrium since one of them is fake. We may assume that F, G and H are heavier. In the third weighing, weigh F and G against K and L. If there is equilibrium, weigh I against J. If there is equilibrium again, the fake coin is H and it is heavy. Otherwise, the fake coin is light, and it is the lighter of I and J. Suppose the third weighing does not result in equilibrium, then F and G must be heavier than K and L. Weigh F against G. If there is equilibrium, the fake coin is K and it is light. Otherwise, the fake coin is heavy, and it is the heavier of F and G. Suppose the first weighing does not result in equilibrium. We may assume that A, B, C, D and E are heavier. In the second weighing, weigh A, B, C, O and P against E, F, G, H, L. If there is equilibrium, the fake coin is among D, M and N. In the third weighing, weigh M against N, and we will know everything. If in the second weighing, E, F, G, H and L are heavier, then the fake coin is E, O or P. In the third weighing, weigh O against P, and we will know everything. Finally, if in the second weighing, A, B, C, O and P are heavier, then the fake coin is among A, B, C and L. In the third weighing, weigh A and B against C and L. We cannot have equilibrium. If C and L are heavier, then C is the fake coin, and it is heavy. Otherwise, weigh A against B, and whichever is heavier is the fake coin.

Problem 8.

We have eleven coins, one of which is fake. Determine whether the fake coin is heavier or lighter than a real coin in two weighings.

Solution:

First weigh three coins against three other coins. If there is equilibrium, all six coins are real. Weigh five of them against the remaining five, and we will know the answer. If there is no equilibrium, the remaining five coins are real. Weigh three of them against the three coins in the heavy pan. If there is equilibrium, the fake coin is light. Otherwise, the fake coin is heavy.

Problem 9.
We have twenty-five coins, two of which are fake. The fake coins have equal weight. Determine whether a fake coin is heavier or in three weighings.

Solution:
Divide the coins into three groups A, B and C, each consisting of eight coins, with 1 left over. Note that A, B and C cannot all have the same total weight or all have different total weights. Hence each may be identified as heavy or light. Their status can be determined in two weighings, for instance, weigh A against B and then B against C. Without loss of generality, we may assume that A is heavy while B and C are light. Now B contains at most one fake coin. Weigh four coins in B against the remaining four coins in B. If there is equilibrium, the fake coins are heavy. otherwise, the fake coins are light.

Problem 10.
We have seven coins, two of which are fake. One is heavier and the other lighter than a real coin. Determine whether their total weight is equal to the total weight of two real coins in four weighings.

Solution:
Let the coins be A, B, C, D, E, F and G. In the first weighing, weigh A against B. If there is equilibrium, both are real. In the second weighing, weigh C against D. If there is equilibrium again, both fake coins are among E, F and G. In the third weighing, weigh ABC against EFG, and we will know the answer. If we do not have equilibrium in the second weighing, one of C and D is fake. We may assume by symmetry that D is heavier. In the third weighing, weigh AB against EF. If there is equilibrium, the fake coins are among C, D and G. If AB are heavier, the fake coins are among D, E and F. If AB are lighter, the fake coins are among C, E and F. A fourth weighing will tell us the answer. Suppose in the first weighing, we do not have equilibrium. Then one of A and B is fake. We may assume by symmetry that A is lighter. In the second weighing, weigh A against C. If there is equilibrium, both are real and B is the heavy fake coin. In the third weighing, weigh DE against FG. Since we cannot have equilibrium, we may assume by symmetry that DE are lighter. Then the fake coins are among B, D and E. If in the second weighing, A is heavier than C, then B is the heavy fake coin and C is the light fake coin. If A is lighter than C, then both B and C are real while A is the light fake coin. In the third weighing, weigh DE against FG. Since we cannot have equilibrium, we may assume that symmetry that DE are lighter. Then the fake coins are among A, F and G. A fourth weighing will tell us the answer.

Problem 11.

There are four coins, some of which may be fake. All fake coins have the same weight, and each is lighter than a real coin. Determine whether the number of fake coins is two in two weighings.

Solution:

Let the coins be A, B, C and D. In the first weighing, weigh AB against CD. Suppose we have equilibrium. In the second weighing, weigh A against B. If we have equilibrium again, then there are no fake coins. Otherwise, exactly one of A and B is fake, and exactly one of C and D is fake. Suppose we do not have equilibrium in the first weighing. In the second weighing, weigh AC against BD. If we have equilibrium, then the number of fake coins is two. Otherwise, the number of fake coins is odd.

Problem 12.

There are sixteen coins. Not all are real, and not all are fake. All fake coins have the same weight, and each is lighter than a real coin. Determine the number of fake coins in nine weighings.

Solution:

Divide the coins into eight pairs and weigh the two coins in the first pair against each other. Suppose there is no equilibrium. Then this is a mixed pair. Weigh every other pair against this pair, and we will know the number of fake coins overall. Suppose the first weighing does not result in equilibrium. Weigh the other pairs one at a time against this pair. At some point we must come across a pair with a different total weight. Weigh the two coins in this pair against each other. We may have a mixed pair, or we can put together a mixed pair by taking one coin from this pair and one coin from the original pair. Now weigh each of the remaining pairs against this mixed pair, and we will know the number of fake coins overall.

Exercises

1. We have four red coins and one blue coin. One of the red coins is fake while the blue coin is known to be real. It is not known whether the fake coin is heavier or lighter than a real coin. Identify the fake coin and determine whether it is heavier or lighter in three weighings.

2. We have five red coins and one blue coin. One of the red coins is fake while the blue coin is known to be real. It is not known whether the fake coin is heavier or lighter than a real coin. Identify the fake coin in two weighings.

3. We have fourteen red coins and one blue coin. One of the red coins is fake while the blue coin is known to be real. It is not known whether the fake coin is heavier or lighter than a real coin. Identify the fake coin in three weighings.

Section 2. Other Problems

In this section, we continue to use the standard balance, but our task is no longer the identification of fake coins or the determination whether they are heavier or lighter than real coins. Our first group of puzzles involves ranking.

Problem 13.
There are sixty-eight coins of different weight. Identify the heaviest and the lightest of the coins in one hundred weighings.

Solution:
Use 34 weighings to compare the 68 coins in pairs, and separate them into the Heavy Group and the Light Group. Clearly the heaviest coin is in the Heavy Group and the lightest coin is in the Light Group. Within each group, we can eliminate the other 33 coins one at a time using 33 weighings. Hence the desired coins can be found using 34+33+33=100 weighings.

Problem 14.
There are sixty-four coins of different weight. Identify the heaviest and the second heaviest coins in sixty-eight weighings.

Solution:
We can determine the heaviest stone in a 6-round knockout tournament. Since 63 stones are eliminated, 63 weighings have been used. Now there are 6 stones which loses directly to the heaviest one, and the second heaviest stone must be among them. Eliminating them one at a time requires 5 additional weighings, bringing the total to exactly 68.

Problem 15.
There are four coins whose weight in grams are four consecutive integers. Construct a complete ranking of the coins by weight in four weighings.

Solution:
Let the coins be A, B, C and D. In the first three weighing, weigh A along with another coin against the other two. We will have equilibrium in exactly one of these three weighings. In the other two weighings, there is a coin which is in the heavier pan both times. It is the heaviest coin of the four. There is also a coin which is in the lighter pan both times. It is the lightest coin of the four. The fourth weighing will complete the ranking.

Problem 16.
There are fifty round coins and fifty-one gold coins. The coins in each group are ordered by weight. All coins have different weights. Identify in seven weighings the coin in the fifty-first place when the one hundred and one coins are ordered by weight.

Solution:

For any positive integer m let $f(m)$ denote the minimum number of weighings required to determine the coin in the m-th place among m ranked gold coins and $m-1$ ranked silver coins. Let $F(m)$ be defined in the same way except with m silver coins also. We claim that these functions have the following properties:

(1) $f(1) = 0$ and $F(1) = 1$;

(2) $f(2k+1) \leq \max\{F(k), f(k+1)\} + 1$;

(3) $f(2k) \leq \max\{F(k), f(k)\} + 1$;

(4) $F(2k+1) \leq f(k+1) + 1$;

(5) $F(2k) \leq F(k) + 1$.

We shall prove only (2), since (3), (4), and (5) can be dealt with in similar manner while (1) is trivial. Let the gold coins be $G_1, G_2, \ldots, G_{2k+1}$ and the silver coins be $S_1, S_2, \ldots, S_{2k}$, listed in descending order of weight. Our strategy is to compare G_{k+1} with S_k in the first weighing. There are two cases.

Case 1. G_{k+1} is heavier than S_k.

Then it is heavier than S_j for $j \geq k$ as well as G_i for $i \geq k+2$. Hence G_{2k+1} is in the $2k$-th place at the lowest, so that G_i cannot be in the $(2k+1)$-st place for $i \leq k+1$. Also, S_{k+1} is lighter than G_i for $i \leq k+1$ as well as S_j for $j \leq k$. Hence S_{k+1} is in the $(2k+2)$-nd place at the highest, so that S_j cannot be in the $(2k+1)$-st place for $j \geq k+1$. Thus we are left with k ranked gold coins and k ranked silver coins, and whichever is in the k-th place among them will be the one in the $(2k+1)$-st place overall. This can be determined in $F(k)$ more weighings.

Case 2. G_{k+1} is lighter than S_j.

As before, we can eliminate G_i for $i \geq k+2$ and S_j for $j \leq k$. The task can be completed in $f(k+1)$ more weighings.

It follows that $f(2k+1) \leq \max\{F(k), f(k+1)\} + 1$, establishing (2). From (2), (3), (4) and (5), we have the following.

(6) For $m \geq 2$, $F(m) \leq n+1$ and $f(m) \leq n+1$, where n is the positive integer such that $2^{n-1} < m \leq 2^n$.

We prove this by induction on m. For $m = 2$, $n = 1$ and it is easily verified that $F(2) = f(2) = 2$. Suppose (6) holds up to $m = 2k-1$. Let $2^{n-1} < 2k \leq 2^n$. Then $2^{n-2} < k \leq 2^{n-1}$. By the induction hypothesis, $F(k) \leq n$ and $f(k) \leq n$. Hence $F(2k) \leq F(k) + 1 \leq n+1$ by (5) and $f(2k) \leq \max\{F(k), f(k)\} + 1 \leq n+1$ by (3). Now let $2^{n-1} < 2k+1 \leq 2^n$. Then $2^{n-2} < k+1 \leq 2^{n-1}$.

By the induction hypothesis, $f(k+1) \leq n$ while $F(k) \leq n$ or $n-1$. It follows that we have $F(2k+1) \leq f(k+1)+1 \leq n+1$ by (4). On the other hand, $f(2k+1) \leq \max\{F(k), f(k+1)\}+1 \leq n+1$ by (2). This completes the inductive argument. Thus we indeed have $f(51) \leq 7$ since $2^5 < 51 \leq 2^6$. A possible strategy is indicated in the proof of (2).

Our next group of puzzles involving balancing objects with a set of tokens.

Problem 17.

You are to choose four tokens such that any object whose weight is an integral number of grams up to some positive integer n, where n is as large as possible, can be balanced by placing it in one pan and a subset of the tokens in the other pan. What should the weights, in grams, of the four tokens be?

Solution:
Since each token is either used or not used, we have 2 choices. Thus the highest possible weight we can balance is $2^4 - 1 = 15$ grams. We subtract 1 to account for the case where none of the tokens are used. We choose the weights of the tokens to be 1, 2, 4 and 8 grams. Note that $1 = 1_2$, $2 = 10_2$, $4 = 100_2$ and $8 = 1000_2$ are just the first 4 powers of 2. To balance an object whose weight is any positive integral number of grams up to 15, convert the number into base 2 and choose the tokens accordingly. For example, since $11 = 1011_2$, we can balance an object of weight 11 grams using tokens of weights $1000_2 = 8$ grams, $10_2 = 2$ grams and $1_2 = 1$ gram.

Problem 18.

You are given 20 tokens such that any object whose weight is an integral number of grams up to 1997 can be balanced by placing it in one pan and a subset of the tokens in the other pan. What is the minimal value of weight, in grams, of the heaviest token if the weights of the tokens are

(a) all integral numbers of grams;

(b) not necessarily integral numbers of grams?

Solution:
Let the weights be $a_1 \leq a_2 \leq \cdots \leq a_{20}$. Clearly, $a_n \leq 2^n$ for all n. In order to minimize a_{20}, we should maximize the others as much as possible.

(a) We may take $a_n = 2^n$ for $1 \leq n \leq 8$, $a_n = 145$ for $9 \leq n \leq 18$ and $a_n = 146$ for $19 \leq n \leq 20$. Using a_1 to a_8, we can balance any integral weight up to 255. By adding the 145's or 146's, we can go all the way up to 1997. Suppose $a_{20} \leq 145$. Then $a_9 + a_{10} + \cdots + a_{20} \leq 1740$ so that $a_1 + a_2 + \cdots + a_8 \geq 257$. This contradicts $a_n \leq 2^n$ for $1 \leq n \leq 8$.

(b) We may take $a_n = 2^n$ for $1 \leq n \leq 8$, $a_n = 145$ for $9 \leq n \leq 12$ and $a_n = 145\frac{1}{4}$ for $13 \leq n \leq 20$. Using a_1 to a_8, we can balance any integral wieght up to 255. By adding the 145's, we can go up to 835. Now we put in four copies of $145\frac{1}{4}$ and free up the 145's for further use. This way, we can go up to 1416. Putting in the other four copies of $145\frac{1}{4}$ allows us to go all the way up to 1997. Suppose $a_{20} < 145\frac{1}{4}$. As in (a), we must still have $a_n = 2^n$ for $1 \leq n \leq 8$. Now $\{a_9, a_{10}, \ldots, a_{20}\}$ must have a subset of size p and total weight $145p + 2$ or two subsets of sizes q and r and total weights $145q + 1$ and $145r + 1$ respectively. Then $p \geq 9$, $q \geq 5$ and $r \geq 5$. Hence at most three of a_9 to a_{20} can be 145, and we will not be able to balance the weight 691.

Problem 19.
Find all sets of tokens such that each weighs an integral number of grams, the total weight is 500 grams and each object weighing an integral number of grams not exceeding 500 grams can be balanced with a *uniquely* subset of the tokens.

Solution:
We may have five hundred tokens of weight 1 gram, two tokens of weight 1 gram plus one hundred and sixty-six tokens of weight 3 grams, or one hundred and sixty-six tokens of weight 1 gram plus two tokens of weight 167 grams. We claim that these are the only possible sets. Suppose such set contains k_1 tokens of weight w_1 grams, k_2 tokens of weight w_2 grams, $\ldots$, plus k_n tokens of weight w_n grams, with $w_1 < w_2 < \cdots < w_n$. Clearly, we must have $w_1 = 1$. Suppose $n > 1$. Since we must be able to balance objects of weight $1, 2, \ldots, w_2 - 1$ grams, we must have $k_1 = w_2 - 1$. For similar reasons, if $n > 2$ then $k_1 + k_2 w_2 = w_3 - 1$ and indeed

$$k_1 + k_2 w_2 + \cdots + k_i w_i = w_{i+1} - 1$$

for $i = 1, \ldots, n-1$. Moreover, $k_1 + k_2 w_2 + \cdots + k_n w_n = 500$. When $i = n-1$, we have $w_n - 1 + k_n w_n = 500$. In other words, $(k_n + 1)w_n = 201 = 3 \times 167$. Since $k_n \geq 1$, the only solutions are $(w_n, k_n) = (1, 500)$, $(3, 166)$ and $(167, 2)$. The first solution clearly implies $n = 1$. The second implies $n = 2$. The third solution implies, with $i = n - 2$, that $w_{n-1} - 1 + k_{n-1}w_{n-1} + 2 \times 167 = 500$, so that $(k_{n-1} + 1)w_{n-1} = 67$. Since $w_{n-1} < w_n = 67$, this implies $w_{n-1} = 1$ and $n = 2$. These correspond to the three sets given earlier.

We conclude with some miscellaneous puzzles.

Problem 20.
There are eight coins, four of which are fake. All fake coins have the same weight, and is less than the weight of a real coin. Pick out two coins of different kinds in two weighings.

Solution:
Weigh four of the coins against the other four. If there is equilibrium, discard one set. Weigh two of the remaining four coins against the other two. If there is equilibrium, take both coins from either pan. Otherwise, take one coin from each pan. Suppose the first weighing does not result in equilibrium. Weigh two of the coins in the heavy pan against the other two in the same pan. If there is equilibrium, take any one of them and any one from the lighter pan in the first weighing. Otherwise, take both coins in the lighter pan.

Problem 21.
There are sixteen coins, two of which are fake. The fake coins have the same weight, which is different from the weight of a real coin. Divide the coins into two groups with equal total weight in three weighings.

Solution:
Label each coin with a four-digit binary number from 0000 to 1111. In the i-th weighing, $1 \leq i \leq 3$, weigh the eight coins with 0 as its i-th digit against the remaining eight coins. If equilibrium is achieved in any of the three weighings, the task is accomplished. Otherwise, the two fake coins are always in the same pan. This means that their binary labels are identical except for the last digit. The task can now be accomplished by putting in one group all coins whose last digits are 0s, and in the other group those whose last digits are 1s.

Problem 22.
There are twenty-two coins, two of which are fake. The fake coins have the same weight, which is different from the weight of a real coin. Divide the coins into two groups with equal total weight in four weighings.

Solution:
Divide the coins into two groups of size eight and two groups of size three. Weigh the two larger groups against each other and the two smaller groups against each other. If we have equilibrium in both cases, or have equilibrium in neither case, the task can be accomplished by combining a larger group with an appropriate smaller group. Suppose equilibrium is achieved only between the smaller groups. Then both fake coins are in one of the two larger groups. Together, they have sixteen coins. By Problem 21, three weighings are sufficient to complete the task. Since we have already performed one such weighing, two more will suffice, yielding a total of four weighings as desired. Suppose equilibrium is achieved only between the larger groups. Then all the coins in them are real. Simply transfer five real coins to each of the smaller groups so that each has eight coins. As before, two more weighings will suffice.

Problem 23.

Nine coins are presented in court as evidence. The judge knows that exactly three of these are real and each weighs 2 grams. Three of the remaining ones are fake coins each weighing 1 gram, while the other three are fake coins each weighing 3 grams. A lawyer claims to know which coins are fake and which are real, and which fake coins are light and which are heavy. How can she prove this to the judge in two weighings?

Solution:
Let A, B and C be the coins each weighing 3 grams, D, E and F be the coins each weighing 2 grams, and G, H and I be the coins each weighing 1 gram. In the first weighing, show that A is equal in weight to G, H and I combined. This is only possible if A indeed weighs 3 grams while each of G, H and I weighs 1 gram. In the second weighing, show that B and C combined are equal in weight to D, E and F combined. This shows that each of B and C weighs 3 grams while each of D, E and F weighs 2 grams.

Problem 24.

Fourteen coins were presented in court as evidence. The judge knows that exactly seven of these are real. The remaining ones are fake coins with the same weight, which is less than the weight of a real coins. A lawyer claims to know which coins are fake and which are real How can she prove this to the judge in three weighings?

Solution:
The lawyer labels the real coins R_1 to R_7 and the fake coins F_1 to F_7. In the first weighing, she puts R_1 in the left pan and F_1 in the right pan. The left pan will be heavier. This demonstrates that R_1 is real and F_1 is fake. In the second weighing, she puts F_1, R_2 and R_3 in the left pan and R_1, F_2 and F_3 in the right pan. The left pan will be heavier. Since the right pan contains a real coin R_1, the left pan must have at least two real coins, which can only be R_2 and R_3. Hence F_2 and F_3 must be fake coins. In the third weighing, she puts F_1, F_2, F_3, R_4, R_5, R_6 and R_7 in the left pan and R_1, R_2, R_3, F_4, F_5, F_6 and F_7 in the right pan. Again the left side will be heavier, and the status of each coin can be correctly deduced.

Exercises

4. There are six coins of different weights, labeled A, B, C, D, E and F in ascending order of weight. It is known that the coins may be divided into two groups of three having the same total weight. Discover such a division in two weighings.

5. Each token weighs a non-integral number of grams. Any object of integral weight from 1 gram to 40 grams can be balanced by placing it in one pan and a subset of the tokens in the same pan. What is the smallest possible number of tokens?

6. We have two green, two red and two white balls. Three balls, one of each color, are of the same weight. The other three are also of the same weight, but the two weights are different. Sort out the heavy balls from the light balls in two weighings.

Section 3. Other Balances

Our first example of a non-standard balance is a one-pan scale which shows the total weight of the coins in the pan.

Problem 25.

We have six coins, one of which is fake. Using a one-pan scale three times, identify the fake coin.

Solution:

Let the coins be A, B, C, D, E and F. In three weighings, we determine the average weight m of C and E, the average weight n of D and F, and the average weight k of B, E and F. If $m = n = k$, the fake coin is A. If $m = n \neq k$, the fake coin is B. If $m \neq n = k$, the fake coin is C. If $k = m \neq n$, the fake coin is D. If $k \neq m \neq n \neq k$, then the fake coin is E or F. This can be distinguished since $2m + n = 3k$ if it is E, and $m + 2n = 3k$ if it is F.

Problem 26.

We have eight coins, one of which is a fake. The weight of a real coin is an integral number of grams. The weight of the fake coin differs from that of a real coin by less than 4 grams. Using a one-pan scale three times, identify the fake coin.

Solution:

Let the coins be A, B, C, D, E, F, G and H. In the first weighing, put D, E, F and H into the pan. If the total weight is an integral multiple of 4, record 0. If it is not, record 1. This digit will be 0 if the fake coin is among A, B, C and G and 1 otherwise. Repeat with C, E, G and H in the second weighing and with B, F, G and H in the third. The three digits recorded will identify the fake coin, as shown in the chart below.

Fake Coin	A	B	C	D	E	F	G	H
First Digit	0	0	0	1	1	1	0	1
Second Digit	0	0	1	0	1	0	1	1
Third Digit	0	1	0	0	0	1	1	1

Our second non-standard balance is a two-pan scale which shows the difference, in grams, between the weights of the two pans.

Problem 27.

Each of four coins weighs an integral number of grams. We can use a faulty two-pan scale which may make a mistake of 1 gram either way in at most one weighing. Determine the weight of each coin in four weighings.

Solution:
Let the coins be A, B, C and D, with respective weights a, b, c and d. In the four weighings, weigh BCD against nothing, AB against CD, AC against BD, and AD against BC. Let the results be $b+c+d = w$, $a+b-c-d = x$, $a-b+c-d = y$ and $a-b-c+d = z$. For now, assume that no mistakes are made. We have $w+x+y+z = 3a$ so that $a = \frac{w+x+y+z}{3}$. Since $y+z = 2a-2b$, we have $b = \frac{2a-(y+z)}{2}$. Similarly, $c = \frac{2a-(z+x)}{2}$ and $d = \frac{2a-(x+y)}{2}$. Suppose now a mistake of 1 gram is possible. If $w + x + y + z$ is a multiple of 3, no mistakes have been made. If it is one more or one less, we know the direction of the mistake. In any case, we can round the total to the nearest multiple of 3 and use it to determine a. Now each of $a - w$, x, y and z has the same parity as $a + b + c + d$, and hence as one another. Whichever has the opposite parity to the other three is where the mistake has been made.

Problem 28.
A team of geologists on a field expedition have taken along 80 cans of food. The cans have different weights, which are recorded on a list. After a while, the labels of the cans have fallen off, and the cook correctly numbers them in ascending order of weight. How can he prove that he has not made any mistakes, using a two-pan scale four times?

Solution:
Introduce an empty can numbered 0. Convert everything into four-digit numbers in base 3, which will run from 0000 to 2222. In the i-th weighing, $1 \le i \le 4$, put in one pan all cans with the digit 0 in the ith position of their base 3 numbering, and in the other pan all cans with the digit 2 there. Thus 27 cans are put on each pan each time. In the first weighing, we obtain the maximum possible difference in weight between any two subsets of 27 cans. Hence the first digits are correct. In the second weighing, we obtain the maximum possible difference in weight between two subsets of 27 cans, consisting of 9 cans whose base 3 expressions start with each of 0, 1 and 2. Hence the second digits are also correct. In the third weighing, we obtain the maximum possible difference in weight between two subsets of 27 cans, consisting of 3 cans whose base 3 expressions start with each of 00, 01, 02, 10, 11, 12, 20, 21 and 22. Hence the third digits are correct too. In the fourth weighing, we obtain the maximum possible difference in weight between two subsets of 27 cans, no two within the same subset having the same first three digits in their base 3 expressions. This proves that the cook has made no mistakes.

The final non-standard balance we consider is invented by Marcus Götz. It has three pans. The one containing the lightest weight will go up. If two pans contain the same weight and the weight of the third one is not less than this, then no pan goes up.

We now use it to solve some standard problems.

Problem 29.
We have four coins, one of which is a fake. It is lighter than a real coin.
Identify the fake coin using the Götz balance once.

Solution:
Leave out a coin and put one of the other three on each pan. If a pan goes
up, it contains the light coin. If none of them goes up, the coin left out is
light.

Problem 30.
We have sixteen coins, one of which is a fake. It is lighter than a real coin.
Identify the fake coin using the Götz balance twice.

Solution:
Divide the coins into four groups of four. In the first weighing, put three
groups into the three pans and leave the fourth out. If a pan goes up, it
contains the light coin. If none of them goes up, the fourth group contains
the light coin. We then continue as in Problem 5.

Problem 31.
We have seven coins, one of which is fake. It is heavier than a real coin.
Identify the fake coin using the Götz balance twice.

Solution:
In the first weighing, divide the coins into two groups of two and one group
of three and put them into the three pans. We can narrow the search down
to three coins (add a normal coin if the heavy coin is in a group of two).
In the second weighing, put two of them into two pans and put two normal
coins into the third. This will identify the heavy coin.

Problem 32.
We have twenty-five coins, one of which is a fake. It is heavier than a real
coin. Identify the fake coin using the Götz balance three times.

Solution:
In the first weighing, divide the coins into two groups of eight and one group
of nine and put them into the three pans. We can narrow the search down
to 9 coins (add a normal coin if the heavy coin is in a group of 8). In the
second weighing, divide them into three groups of three. Put two of them
into two pans, and put four normal coins into the third. We can now narrow
the search down to three coins. We then continue as in Problem 7.

Problem 33.
We have four coins, two of which are fake. Each is lighter than a real coin.
Identify the two fake coins using the Götz balance twice.

Solution:
Let the coins be A, B, C and D. In the first weighing, put A in the first pan, B in the second pan, and CD in the third pan. The third one cannot go up. If one pan goes up, we may assume by symmetry that it is the first one. Then A is a fake coin. In the second weighing, take off A and move D to the first pan. Whichever pan goes up contains the other fake coin. Suppose in the first weighing, none of the pans goes up. Then either A and B are fake or C and D are fake. In the second weighing, switch B and D. Either the first or the second pan must go up. If the first pan goes up, the fake coins are A and B. If the second pan goes up, the fake coins are C and D.

Problem 34.
We have six coins, two of which are fake. Each is heavier than a real coin. Identify the two fake coins using the Götz balance twice.

Solution:
Let the coins be A, B, C, D, E and F. In the first weighing, put AB in the first pan, CD in the second pan and EF in the third pan. Suppose none of them goes up. This means that both fake coins are in the same pan. In the second weighing, take one coin off each of two pans. If either of these goes up, the other contains both fake coins originally. If neither of them goes up, the third pan, which cannot go up, contains both fake coins. Suppose in the first weighing, one of them goes up. By symmetry, we may assume that it is the first. This means that one of the fake coins is C or D, and the other is E or F. In the second weighing, move B to the second pan, D to the third pan and F to the first pan. If none of them goes up, the fake coins are D and E. If the first pan goes up again, the fake coins are C and E. If the second pan goes up, the fake coins are D and F. If the third pan goes up, the fake coins are C and F.

Exercises

7. We have six coins two of which are fake. Each fake coin weighs 1 gram more than a real coin. We can use a faulty standard balance which remains in equilibrium if the difference between the weights of the two pans is less than 2 grams. Identify both fake coins in four weighings.

8. We have six coins, two of which are fake. Each is lighter than aa real coin. Identify the two fake coins using the Götz balance three times.

9. We have nine coins, two of which are fake. Each is heavier than a real coin. Identify the two fake coins using the Götz balance three times.

Bibliography

[1] Hess, R. I. *Counterfeit Coins*, exchange item in the seventh Gathering for Gardner, Atlanta, 2006.

[2] Kordemsky, B. *Moscow Puzzles*, Dover, Mineola, 1987.

[3] Taylor, P. J. *International Mathematics Tournament of the Towns: 1980–1984*, AMT, Canberra, 1993.

[4] Taylor, P. J. *International Mathematics Tournament of the Towns: 1984–1989*, AMT, Canberra, 1992.

[5] Taylor, P. J. *International Mathematics Tournament of the Towns: 1989–1993*, AMT, Canberra, 1994.

[6] Storozhev, A. M. and Taylor, P. J. *International Mathematics Tournament of the Towns: 1993–1997*, AMT, Canberra, 1983.

[7] Storozhev, A. M. *International Mathematics Tournament of the Towns : 1997–2002*, AMT, Canberra, 2006.

[8] Liu, A. and Taylor, P. J. *International Mathematics Tournament of the Towns : 2002–2007*, AMT, Canberra, 2009.

Solution to Exercises

1. We choose base 3 for the same reason as in Problem 2. Since we have two weighings, we look at two-digits numbers, of which there are nine. They are 11 and the four complementary pairs (00,22), (01,21), (12,10) and (20,02). The fake coin is one of four, and it may be heavy or light. Thus there are 8 possible outcomes in our investigation. We assign 11 to the blue coin. We now construct the Primary Codes for the red coins. Among them, we want two with 0 as their first digits, one with 1 as their first digits, and one with 2 as their first digits. The same applies to their second digits. Thus the first number in each of the pairs above is taken to be the Primary Code of a red coin. For now, assume that the fake coin is heavy. In the first weighing, place the two coins with 0 as the first digits of their Primary Codes in one pan and those with 1 as the first digits of their Primary Codes in other pan. In the second weighing, place the two coins with 0 as the second digits of their Primary Codes in one pan and those with 1 as the second digits of their Primary Codes in other pan. Thus we can determine the Primary Code of the fake coin, on the assumption that it is heavy. We then continue as in Problem 2.

2. Let the blue coin be A and the red coins be B, C, D, E and F. In the first weighing, weigh AC against BE. In the second weighing, weigh AD against BC. If we have equilibrium both times, the fake coin is F. If we have equilibrium only in the second weighing, the fake coin is E. If we have equilibrium only in the first weighing, the fake coin is D. If we do not have equilibrium but the balance tilts in opposite directions in the two weighings, the fake coin is C. If it tilts in the same direction, the fake coin is B.

3. Let the blue coin be A and the red coins be B, C, D, E, F, G, H, I, J, K, L, M, N and O. We want the following results from the three weighings. If we have equilibrium all three times, the fake coin is O. If we have equilibrium twice, the fake coin is L, M or N according to which weighing does not result in equilibrium. If we have equilibrium only in the first weighing, the fake coin is F or I. If we have equilibrium only in the second weighing, the fake coin is G or J. If we have equilibrium only in the third weighing, the fake coin is H or K. The two alternatives are decided according to whether the balance tilts in the same direction or in opposite directions in the other two weighings. Suppose we do not have equilibrium at all. If the balance tilts in the same direction all three times, the fake coin is B. If it tilts in the different direction only in the first weighing, the fake coin is C. If it tilts in the different direction only in the second weighing, the fake coin is D. If it tilts in the different direction only in the third weighing, the fake coin is E. This allows us to construct the chart below, on the assumption that the fake coin is heavy. We could assume that it is light, and obtain a chart with all inequalities reversed.

	B	C	D	E	F	G	H	I	J	K	L	M	N	O
1st	<	<	>	>	=	<	<	=	>	>	<	=	=	=
2nd	<	>	<	>	>	=	<	>	=	<	=	<	=	=
3rd	<	>	>	<	>	<	=	<	<	=	=	=	>	=

The chart now allows us to construct the weighing algorithm. In the first weigh, we weigh ADEJK against BCGHL. In the second weighing, we weigh ACEFI against BDHKM. In the third weighing, we weigh ACDFN against BEGIJ.

4. The only possible groupings are ABF against CDE, ACF against BDE, ADF against BCE, AEF against BCD and ADE against BCF. First weigh ADF against BCE. If there is equilibrium, the task is accomplished. If ADF are heavier, then AEF will be heavier than BCD. Then we weigh ACF against BDE. If there is equilibrium, the task is accomplished. If ACF are heavier, then BCF will be heavier than ADE. Hence ABF must have the same total weight as CDE.

If in the first weighing ADF are lighter, then ACF will be lighter than BDE, ABF will be lighter than CDE and ADE will be lighter than BCF. Hence AEF must balance BCD.

5. Seven tokens are sufficient, with weights 0.5, 0.5, 1.5, 2.5, 5.5, 10.5 and 21.5 grams respectively. Each is obtained from the preceding one by doubling and then alternately subtracting and adding 0.5 grams. It is routine to verify that every object with integer weight up to 41 grams can be balanced. Suppose we only have six tokens. Put one of them aside. Then there are $2^5 = 32$ sets formed from the other five tokens. We associate each set with a companion set formed by adding the token set aside. Since this token has a non-integer weight, at most one set from each associated pair can balance an object with integer weight. Since $32 < 40$, six tokens are not sufficient.

6. First, weigh one green and one red ball against one white and the other red ball. If there is equilibrium, we have a heavy ball and a light ball on each side. Now weigh the remaining green and white balls against each other. One of them will be heavy and the other light. We will then know whether the green ball in the first weighing is heavy or light. Whichever it is, the red ball that is on its side will be the opposite. Suppose the first weighing does not result in equilibrium. Then the red ball on the heavy side is heavy. As for the green and white balls involved in the first weighing, both may be heavy, both may be light, or the one on the heavy side is heavy while the other is light. Now weigh these two balls against the two red balls, and we can distinguish among the three cases.

7. Let the coins be A, B, C, D, E and F. In the first two weighings, weigh ABC against DEF, and ABD against CEF. Suppose neither weighing results in equilibrium. Then the fake coins are either A and B or E and F, and the results of both weighing will show this with consistence. Suppose equilibrium is achieved only once. We may assume that ABC are heavier than DEF. Then C is clearly fake, and the other one is A or B. In the third weighing, weigh BC against AD. BC cannot be lighter than AD. If BC are heavier, then they are the fake coins. Otherwise, A and C are the fake coins. Finally, suppose we have equilibrium in the first two weighings. Then C and D are either both real or both fake. In the third weighing, weigh ACD against BEF. If ACD are heavier, then C and D are the fake coins. If ACD are lighter, then B and either E or F are the fake coins. If equilibrium is achieved, then A and either E or F are the fake coins. In the fourth weighing, weigh AE against BF. If one pan is heavier, then it contains the two fake coins. Otherwise, we have one fake coin in each pan, which also allows us to identify both fake coins.

8. Let the coins be A, B, C, D, E and F. In the first weighing, put A in the first pan, B in the second pan and CD in the third pan. The third one cannot go up. If either of the first two goes up, we may assume by symmetry that it is the first one. Then A is a fake coin. In the second weighing, take A off and move D to the first pan. If any pan goes up, it contains the other fake coin. Suppose none of them goes up. This means that the other fake coin is E or F. In the third weighing, replace B by E and C by F. Whichever of the second and the third pans goes up contains the other fake coin. Suppose in the first weighing, none of the pans goes up. Then A and B are either both fake or both real. In the second weighing, replace A by E and B by F. The third pan cannot go up. If either of the others goes up, we may assume by symmetry that it is the first one. Then E is a fake coin. In the third weighing, put C, D and F into the three pans. Whichever goes up contains the other fake coin. Suppose none of the pans goes up in the second weighing. This means that the fake coins are A and B, C and D, or E and F. In the third weighing, replace E by A. If the first pan goes up, the fake coins are A and B. If the second pan goes up, the fake coins are E and F. If none of them goes up, the fake coins are C and D.

9. We give a non-adaptive solution. Let the coins be A, B, C, D, E, F, G, H and I. In the first weighing, we put ABC, DEF and GHI into the three pans respectively. In the second weighing, we put ADG, BEH and CFI into the three pans respectively. In the third weighing, we put AEI, BFG and CDH into the three pans respectively. In any weighing, if no pans go up, this means that the two fake coins are in different pans. This can only happen once. Suppose it does happen. We may assume by symmetry that it happens in the first weighing, and that the first pan goes up in the second weighing. In the third weighing, if the first pan goes up, the fake coins are B and C. If the second pan goes up, the fake coins are H and I. If the third pan goes up, the fake coins are E and F. Henceforth, suppose that some pan always goes up. We may assume by symmetry that the first pan goes up in both the first and the second weighing. Then the fake coins are either E and I or F and H. Since no pans go up in the third weighing, they cannot be E and I. It follows that the fake coins are F and H.

Chapter Seven: Graph Theory

Section 1. Basic Concepts

A **graph** is a structure of dots and lines. The dots are called **vertices** and the lines are called **edges**. Each edge joins two vertices, and is **incident** with them. If the two vertices are the same, the edge is called a **loop**. Two different edges may join the same pair of vertices. They are then called **multiple edges**. Unless otherwise stated, we assume that our graphs have no loops or multiple edges.

A graph with no edges is called an *empty graph*. A graph with all possible edges is called a **complete graph**. A **bipartite** graph is one where the vertices are divided into two disjoint subsets, and each edge joins one vertex from each subset.

A **subgraph** is a subset of the vertices and edges of the graph. If all the vertices of the graph are in the subgraph, then we have a **spanning subgraph**.

The **degree** of a vertex is the number of edges incident with it. If there is a loop at a vertex, then it contributes 2 to the degree of that vertex. A vertex with even degree is called an **even** vertex, and a vertex with odd degree is called an **odd** vertex.

The Degree Theorem.
The sum of the degrees of the vertices in a graph is equal to twice the number of edges in the graph.

Proof:
This is because each edge contributes 2 to the sum of the degrees of the vertices.

The Parity Theorem.
The sum of the degrees of all vertices of a graph is even.

Proof:
This is follows immediately from the Degree Theorem.

Corollary.
The number of odd vertices in a graph is even.

Proof:
The sum of the degrees of the even vertices is clearly even. It follows from the Parity Theorem that the sum of the degrees of the odd vertices is also even. Hence we must have an even number of odd vertices.

© Springer International Publishing AG 2018
A. Liu, *S.M.A.R.T. Circle Minicourses*, Springer Texts in
Education, https://doi.org/10.1007/978-3-319-71743-2_7

A classic example involving graphs is the analysis of the puzzle called Instant Insanity. It consists of four unit cubes with faced painted in four colors, R for red, Y for yellow, B for blue and G for green. One version of this puzzle is given in Figure 7.1

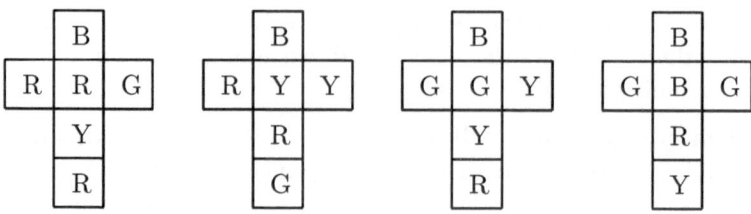

Figure 7.1

The task is to put the four cubes into a $1 \times 1 \times 4$ stack so that on each of the four 1×4 walls, all four colors are present.

In constructing a graph to represent the data of this puzzle, it is obvious that we should have four vertices, each representing one of the colors. However, how should the edges be constructed, and what would they represent?

The key observation is that if one face of a cube appears in one of the four walls, then the opposite face will also appear, on the opposite wall. If it does not appear on any wall, which means it is either on top of at the bottom after the cube has been placed, the opposite face will also not appear.

So we represent a pair of opposite faces of a cube by an edge. The vertices they join represent the colors of these faces. Hence each cube is represented by three edges, and we label the edges 1, 2, 3 or 4 according to whether they come from the first, the second, the third or the fourth cube. Note that our graph may have loops or multiple edges. Figure 7.2 is the graph for the given version of the puzzle.

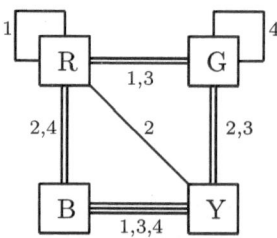

Figure 7.2

We now turn our attention to the desired configuration. Each pair of opposite faces is represented by a subgraph consisting of four edges with different labels, such that each vertex has degree 2. We seek two such subgraphs which are disjoint. It is easy to check that neither of them can have loops or multiple edges. In fact, they have to be the two subgraphs shown in Figure 7.3. These can be used to put together the desired configuration.

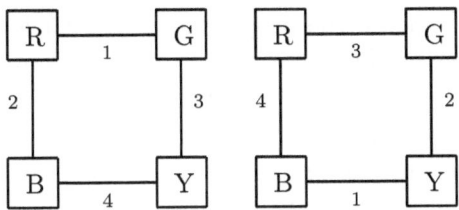

Figure 7.3

This example illustrates the versatility of graphs, in that the vertices and edges can represent almost anything. The most common scenario modeled by a graph is a transportation network, where the vertices represent communities while the edges represent roads. Many problems involving moving around a graph, going from vertices to vertices along edges.

There are many others examples in which the edges are labeled or colored.

Problem 1.
The game Minesweeper is played on a 10×10 chessboard. Each square either contains a bomb or is vacant. On each vacant square is recorded the number of bombs in the neighboring squares along a row, a column or a diagonal. Then all the bombs are removed, and new bombs are placed in all squares which were previously vacant, and the numbers of neighbors with bombs are recorded as before. Can the sum of all numbers on the board now be greater than the sum of all numbers on the board before?

Solution:
Construct a graph with 100 vertices representing the squares, and join two vertices by an edge if they represent neighboring squares. An edge is painted red if it joins a bomb square and vacant square, and it will contribute 1 to the number recorded on the vacant square. An edge which joins two bomb squares or two vacant squares is unpainted. Hence the sum of the recorded numbers is equal to the number of red edges. After the transformation of the board, a red edge remains a red edge, and an unpainted edge remains an unpainted edge. Hence the sum of all the numbers on the board before must be equal to the sum of all the numbers on the board after.

Here is another classic problem.

Problem 2.
There are six people at a party. Every two of them either love each other or hate each other. Prove that there exist two trios of people in which all three pairwise relations are of the same kind.

Solution:
Construct a graph with six vertices representing the six people. Two vertices are joined by a blue edge if the people they represent love each other, and by a red edge otherwise. There are $\binom{6}{3} = 20$ triangles overall. Let x be the number of those that are monochromatic.

Define an arrow as two edges incident with a common vertex. We count the number of monochromatic triangles in two ways. Each triangle defines three arrows. For a monochromatic triangle, all three arrows are monochromatic. For a non-monochromatic triangle, exactly one of the three arrows is monochromatic. Hence the total is $3x + (20 - x) = 2x + 20$. Now five edges are incident with each vertex, which defines $\binom{5}{2} = 10$ arrows. According to whether the color split of these five edges is 5:0, 4:1 or 3:2, the number of these ten arrows that are monochromatic is 10, 6 or 4, respectively. Hence $2x + 20 \geq 6 \times 4$, which yields $x \geq 2$.

This problem generalizes to a deep concept known as Ramsey Theory.

A **path** is a sequence of edges such that two consecutive edges in the sequence share a common vertex. If the initial vertex of a path coincides with the final vertex, then the path closes up to a **cycle**. A path or a cycle with n edges has **length** n, and is called an n-path or an n-cycle respectively. Thus an arrow in the last problem is just a 2-path. In graphs with loops and multiple edges, a loop is a 1-cycle while two double edges constitute a 2-cycle.

Problem 3.
At a party, each person knows at least three other people. Prove that an even number of them, at least four, can sit at a round table such that each knows both neighbors.

Solution:
Construct a graph with the vertices representing the people, and join two vertices by an edge if they represent people who know each other. Consider the longest path in the graph. Let the first vertex be a. Then all the vertices adjacent to a must be in this path, as otherwise any missing one could be put in front of a to form a longer path. Let b, c and d be the first three vertices down the line which are adjacent to a. Suppose there are an odd number of vertices between b and c. Then we can put the people represented by the vertices in the path b to c at a round table and insert a between b and c. This will meet the condition of the problem.

Similarly, if there are an odd number of vertices c and d, the condition can also be met. Finally, if there is an even number, possibly 0, of vertices both between b and c, and between c and d, then there are an odd number of vertices between b and d.

Problem 4.
Prove that in any group of 50 people there are always two who have an even number of common acquaintances within the group.

Solution:
We construct a graph with the 50 vertices representing the people and the edges representing acquaintance. Suppose to the contrary that every two vertices are joined by an odd number of 2-paths. Consider a particular vertex v. Divide the remaining vertices into the set A of its neighbors and the set B of its non-neighbors. If a vertex a in A is joined by a 2-path to v, the middle vertex must be in A. Hence a must have odd degree within A, so that A has an even number of vertices. It follows that v has even degree, as do all other vertices since any can play the role of v. It follows that each a in A is joined to an even number of vertices in B, so that the total number of edges between A and B is even. Now if a vertex b in B is joined by a 2-path to v, the middle vertex must again be in A. Hence it is joined to an odd number of them. It follows that B also has an even number of vertices, so that the total number of vertices is odd. This is a contradiction since there are 50 vertices.

A graph is said to be **connected** if for any two vertices, there exists a path with them as the initial and the final vertices. If a graph is not connected, each connected piece is called a **component** of the graph. A connected graph has only one component.

Problem 5.
Anna and Boris play a game on an archipelago with 2009 islands. Some pairs of islands are connected by boats which run both ways. Anna starts by choosing one of the islands, and turns alternate thereafter. In each turn, the player chooses the next island to be visited. It must be connected by boat to the island where they currently are, and it must not have been visited before. The player who cannot make such a choice loses the game. Prove that Anna has a winning strategy on every possible archipelago.

Solution:
We construct a graph, with the vertices representing the islands and the edges representing connecting routes. The graph may have one or more connected components. Since the total number of vertices is odd, there must be a connected component with an odd number of vertices. Anna chooses from this component the largest set of independent edges, that is, edges no two of which have a common endpoint, and paints them red.

Since the number of vertices is odd, there is at least one vertex which is not incident with a red edge. Anna will start the tour there. Suppose Boris has a move. It must take the tour to a vertex incident with a red edge, as otherwise, Anna could have color one more edge red. Anna simply continue the tour by following that red edge. If Boris continues to go to vertices incident with red edges, Anna will always have a ready response. Suppose somehow Boris manages to get to a vertex not incident with a red edge. Consider the tour so far. Both the starting and the finishing vertices are not incident with red edges. In between, the edges are alternately red and unpainted. If Anna interchanges the red and unpainted edges along this tour, she could have obtained a larger independent set of edges. This contradiction shows that Boris could never get to a vertex not incident with red edges, so that Anna always wins if she follows the above strategy.

Problem 6.
A country has more than 101 towns. Its capital is connected by direct flights with 100 of the towns. Each town other than the capital is connected by direct flights with 10 other towns which may include the capital. At the moment, it is possible to travel by air from any town to any other town, changing planes if necessary. An economy drive requires closing down at least 50 of the flights from the capital. Prove that this can be done so that it is still possible to travel by air from any town to any other town, changing planes if necessary.

Solution:
Construct a graph G in which each city is represented by a vertex and each direct flight by an edge. There are simply too many varieties of this graph for us to handle comfortably. Instead of deciding which flights from the capital we should close down, we close down all of them and consider which flights from the capital we should reconnect. Now the situation facing us is much simpler. Let G' be the subgraph of G by removing M, the vertex representing the capital, and all edges incident with it. By hypothesis, G is a connected graph, but G' may consist of a number of components. However, each component must contain at least one vertex connected to M in G. In G', such vertices have degree 9 while all others have degree 10. By the Corollary to the Parity Theorem, each component must have an even number of vertices with odd degrees. Hence at least two vertices in each component are connected to M in G. Since M has degree 100, the number of components in G' is at most 50. Hence we can make G' a connected subgraph by restoring M and one edge connecting it to each component of G'.

Problem 7.

There are 100 points on the plane, no three on a line. All 4950 pairwise distances between two points have been recorded. Some of them have been erased. What is the maximum number of erased records such that they can always be restored from the remaining records?

Solution:

Suppose at most 96 records are erased. Construct a graph with 100 vertices representing the 100 points. Two vertices are joined by an edge if the record of the distance between the two points they represent is erased. The graph has at most 96 edges, and therefore at least 4 components. Take four vertices a, b, c and d, one from each component, representing the points A, B, C and D. The pairwise distances between these four points are on record, so that their relative positions can be determined. For any other point P, the vertex which represents it is in the same component with only one of a, b, c and d. Hence the distance between P and three of the points A, B, C and D are on record. This is enough to determine the position of P relative to the points A, B, C and D. It follows that all erased records may be restored. Suppose 97 records are erased. All of them may be associated with a point A so that we only know the distances AB and AC, where B and C are 2 of the other 99 points. A does not lie on BC as no three of the 100 points lie on a line. Now we cannot determine whether A is on one side or the other side of the line BC.

The classic Seven-Bridge Problem is based on the concepts of parity and connectivity. Figure 7.4 shows a stylized portion of the River Pregel in Königsburg, a Prussian town at the time. It is now the Russian town Kaliningrad. The two islands are linked to the shores and to each other via seven bridges.

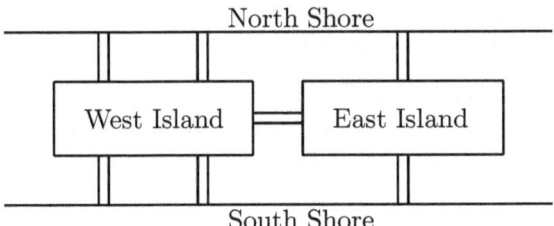

Figure 7.4

It is said that the great Swiss mathematician Euler, who lived there, liked to take a stroll over the islands. He discovered that he could not cross each of the seven bridges exactly once, even if he did not have to return to his starting point.

A simple explanation is offered by graph theory. In the graph in Figure 7.5, each vertex represents one of the four land masses, and each edge represents one of the seven bridges. There are two pairs of double edges. The degrees of the vertices are 5. 3. 3. and 3.

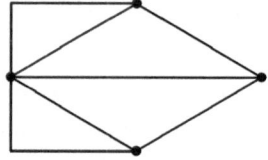

Figure 7.5

An **Eulerian** path is a path which includes all edges of a graph. An **Eulerian** cycle is a cycle which includes all edges of a graph. If we write down the two vertices of each edge along the Eulerian path, each vertex is written down an even number of times except for the initial and the final vertices. When these two vertices coincide in an Eulerian cycle, there will be no exception.

It follows that a necessary condition for a connected graph to have an Eulerian cycle is that is has no odd vertex. This condition is also sufficient. Start from any vertex and keep going until we are stuck. Since every other vertex is an even vertex, and each visit to any vertex uses up two edges incident with it, we cannot get stuck anywhere except back where we have started. If this cycle includes all the edges, we have an Eulerian cycle. Suppose this is not the case. Since the graph is connected, we can take a detour from some vertex on this cycle without using edges already in the cycle, and keep going until we return to that vertex. The detour can then be incorporated, yielding an expanded cycle. Similarly, a connected graph has an Eulerian path if and only if it has at most two odd vertices.

A companion concept to an Eulerian cycle is a Hamiltonian cycle, which passes through every vertex exactly once. While the problem of Eulerian cycles is completely solved, the problem of Hamiltonian cycles is wide open.

Problem 8.
The graph in Figure 7.6 is known as Petersen's graph. Prove that it has no Hamiltonian cycles, but for any vertex, there is a cycle which passes through the remaining nine vertices.

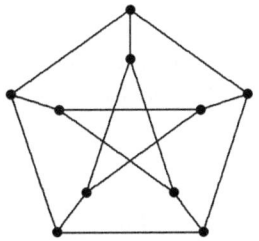

Figure 7.6

Solution:
Assuming to the contrary that Petersen's graph has a Hamiltonian cycle. As drawn, it has five "outer" vertices and five "inner" vertices. If we go from one outer vertex to an inner vertex, we cannot go immediately to another outer vertex. Similarly, we cannot have an outer vertex wedged between two inner vertices along the cycle. Hence the Hamilton cycle must be of the form OOOOOIIIII or OOOIIIOOII, where O stands for an outer vertex and I for an inner vertex. It is easy to verify that in either case, we cannot complete the cycle. Now all the vertices in Petersen's graph are symmetric to one another. Thus we may assume that the top vertex is omitted. Figure 7.7 shows a cycle passing through the remaining nine vertices.

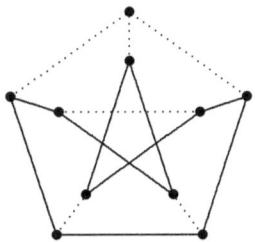

Figure 7.7

Exercises

1. On the first day, each of twenty teams plays against one of the other team. On the second day, each team again plays against one other team, but not the same team it has played on the first day. Prove that there exist ten teams no two of which have yet played each other.

2. An $n \times n \times n$ cube is constructed with n^3 unit cubes each of which is white or black. Each unit cube shares a common face with exactly three other unit cubes of opposite colors. For which positive integers n is this possible?

3. A village consists of 9 blocks in a 3×3 configuration. Each block is a square of side length 1. There are four streets each of length 3, two along edges of the village and two separating columns of blocks. Similarly, there are four avenues each of length 3, two along edges of the village and two separating rows of blocks. Starting from any intersection, what is the smallest distance the Village Dustman must travel along the roads, if he must sweep every section of roads at least once and finish at the starting point?

Section 2. Trees

Two of the basic concepts in graph theory covered in Section 1 are connectivity and cycles. Together, they form the basis of a special kind of graphs. A **tree** is defined as a graph such that (1) it is connected and (2) it has no cycles. Thus every path is a tree. Another example of a tree consists of a vertex joined to all the others. It is called a **star** and an n-star if it has n edges. Note that a 2-path is also a 2-star.

Connectivity Theorem.
In a connected graph with V vertices and E edges, $E \geq V - 1$.

Proof:
We remove all the edges, so that each of the V vertices constitutes a component of the empty subgraph. As each edge is restored, the number of components can reduce by at most 1. Since we need to reduce the number of components by $V - 1$, we must have at least $V - 1$ edges.

Every tree has an additional property (3) which is stated as follows.

The Tree Formula.
Let V and E denote the numbers vertices and edges of a tree respectively. Then $E = V - 1$.

Proof:
We use a physical model of the graph. Let the vertices be beads and the edges be threads linking the beads. If we pick up one of the beads, the whole structure comes up because the graph is connected. Each bead apart from the one we are holding is dangling at the end of exactly one thread. This is because the graph has no cycles. The Tree Formula follows immediately.

Corollary.
A tree has at least two vertices of degree 1.

Proof:
Let the number of vertices of a tree be V. Then it has $V - 1$ edges by the Tree Formula. The total degree of all the vertices is $2V - 2$ by the Degree Theorem. Hence the average degree is less than 2, so that there is at least one vertex of degree 1. By the Corollary to the Parity Theorem, the number of such vertices must be even. Hence there are at least two vertices of degree 1.

Such a vertex is called an **terminal** vertex. The Tree Formula reminds us very much of the Connectivity Theorem. Thus a tree may be regarded as a minimally connected graph.

A graph which satisfies only one of (1), (2) and (3) is not necessarily a tree. Figure 7.8 shows three counterexamples.

Figure 7.8

The first one has no cycles, but is not connected and does not satisfy the Tree Formula. Such a graph is called a **forest**. The second one is connected, but has cycles and does not satisfy the Tree Formula. The third one satisfies the Tree Formula, but is not connected and has cycles. However, if a graph satisfies two of (1), (2) and (3), it must also satisfy the third one. We have already seen that (1) and (2) imply (3).

Suppose a graph satisfies (2) and (3) but not (1). Then it has no cycle but is not connected. Add a new edge between two vertices in different components. No cycle is created, and the number of components reduces by 1. Continuing in this manner, we eventually obtain a graph which satisfies (1) and (2). Hence it must now satisfies (3). However, it already satisfies (3) before new edges are introduced. Thus we have a contradiction.

Suppose a graph satisfy (3) and (1) but not (2). Then it is connected but also has cycles. Delete an edge on any cycle. This will not disconnect the graph, and the number of cycles is reduced by at least 1. Continuing in this manner, we eventually obtain a graph which satisfies (1) and (2). Hence it must now satisfies (3). However, it already satisfies (3) before edges are deleted. Thus we also have a contradiction.

Problem 9.
There are m identical chocolate bars which are to be shared equally among n children. Each bar is either given to one child or broken into two pieces and given to two children. For a given value of m, what are the values of n for which the task is possible?

Solution:
We first prove that if $m \geq n$, the task is always possible. Put the m bars end to end in a row and cut the row into n equal portions. Since each portion consists of at least one complete bar or an equivalent among, no bar is shared by three or more children. Suppose $m < n$ and the task is possible. Now a fair share is less than a whole bar, so that each is shared between exactly two children. Construct a graph with n vertices representing the children. Two vertices are joined by an edge if and only if the children they represent share a bar. There will be m edges. The graph contains no cycles because a cycle with k vertices and k edges represent k children sharing themselves at least k bars. Hence each fair share is at least 1 complete bar, which is a contradiction. It follows that the graph is a forest. In a tree with k vertices, there are exactly $k-1$ edges. Hence a fair share is $\frac{k-1}{k}$ of a bar. It follows that if there are two or more trees in the forest, each tree must have the same number of vertices.

For a given value of m, the values of n for which the task is possible are those satisfying $n \leq m$ and those satisfying $n = m + k$ where k is any positive divisor of m.

Problem 10.
Seven children are coming to a party. They are four more such that either they will all come or none of these four will come. The host buys 77 pieces of chocolate, so that a fair sharing is possible whether seven or eleven children come. To save distribution time, she puts them into bags, not necessarily the same number of pieces in each. When the children come, each will get a number of bags in a fair sharing. What is the minimum number of bags?

Solution:
The host may prepare 17 bags, containing 7, 7, 7, 7, 7, 7, 7, 4, 4, 4, 4, 3, 3, 3, 1, 1, and 1 pieces respectively. When eleven children come, the distribution is $7=7=7=7=7=7=7=4+3=4+3=4+3=4+1+1+1$. When seven children come, the distribution is $7+4=7+4=7+4=7+4=7+3+1=7+3+1=7+3+1$. We now prove that 17 is minimum. Construct a bipartite graph with 7 vertices on one side representing the seven children who are sharing, and 11 vertices on the other side representing the eleven children who are sharing. Every edge represents a bag of pieces of chocolate. An edge joins two vertices if that bag is given to the two persons in the respective scenarios. We claim that the graph is connected. If not, some of the chocolate pieces are shared equally among m children in the first scenario and by n children in the second scenario, where $m < 7$ and $n < 11$. Then the number of chocolate pieces involved is $11m = 7n$. Since 7 and 11 are relatively prime, we cannot have $m < 7$ or $n < 11$. Hence the graph must be connected. Thus the minimum number of edges in this graph with 18 vertices is 17.

Problem 11.
Each of six people has a piece of juicy gossip, and would like to spread it among the others. Various phone calls take place. In each call, both parties tell the other everything they know. How many calls does it take for all six people to know all pieces of gossip?

Solution:
We first prove that 8 calls are sufficient. Let the people be A, A', B, B', C and D. In the first round, A' calls A and B' calls B. In the next two rounds, A, B, C and D exchange all information with A calling B and C calling D, and then A calling D and B calling C. In the final round, A calls A' and B calls B'. We now suppose that 7 calls are still sufficient. Construct a graph G with six vertices representing the people A, B, C, D, E and F, and seven edges representing the calls. We that a person is well-informed when all six pieces of gossip are known to this person. Now the subgraph H consisting of the edges representing the first four calls is not connected. Hence nobody is well-informed after the fourth call.

On the other hand, all are supposed to become well-informed after three more calls. Hence the last 3 calls must be between disjoint pairs of people, say between A and D, between B and E, and between C and F. When these three edges are added to H, we must have a connected subgraph of G. It follows that H has exactly two components, one consisting of say A, B and C, and the other D, E and F. Moreover, of the four edges in H, exactly two are in each component. We may assume that A is not involved in the second of the two calls among A, B and C. Then A does not know either B's or C's piece of gossip. After A calls D, A is still not well-informed. This is a contradiction.

Problem 12.
The 52 playing cards of a standard deck are arranged face-up in a row. The order is visible to the audience which includes an assistant of the magician, but not to the magician himself. The assistant may name two cards and state how many other cards lie in between them. What is the minimum number of statements that the assistant must make so that the magician can deduce the relative order of the cards?

Solution:
The minimum number of statements is 34. Construct a graph with 52 vertices representing the cards, numbering them from 1 to 52 in the given order. An edge joins two vertices if a statement is made by the assistant about the two cards represented by these vertices. We first prove that 34 statements are necessary. Clearly, the graph can have at most one isolated vertex. Every other component must contain at least two edges, as otherwise the two cards represented by the vertices joined by the sole edge can be switched around. Thus each component with at least two vertices must have at least three. Minimizing the number of edges is equivalent to maximizing the number of components and avoiding cycles within the components. Thus our optimal graph is a forest consisting of one isolated vertex and 17 2-paths, so that the minimum number of edges is 34. Hence 34 statements are necessary. We now prove that 34 statements are also sufficient. Figure 7.9 shows such a forest with 34 edges, each labeled with two numbers. The first number denotes the order in which the statement is made, and the second number is what is made in the statement. After statement 1, the magician will know that cards 1 and 52 are at the ends of the arrangment. After statement 2, he knows that card 3 is separated from card 1 by one other card. After statement 3, either card 2 is next to card 1 and card 51 is next to card 52, or vice versa. In the latter case, card 49 is on the other side of card 3 by statement 4. It is however no longer possible for 45 cards to go between cards 4 and 50, as declared in statement 5. It follows that the first case prevails.

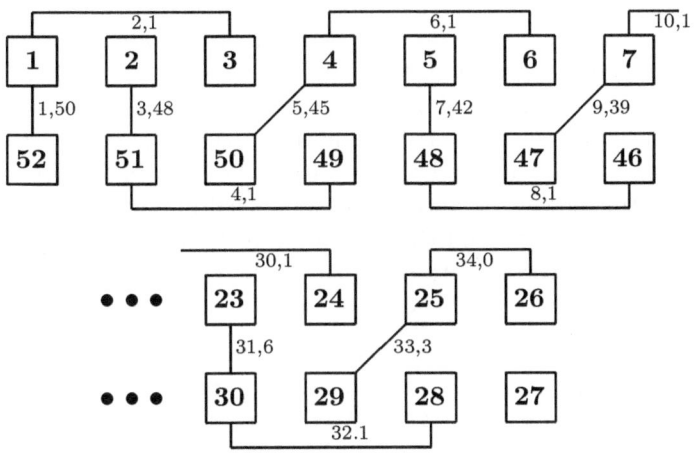

Figure 7.9

Statement 5 itself presents an ambiguity in whether card 4 or card 50 is next
to card 51, but as before, consideration of statement 7 will settle the issue.
Continuing in this manner, we can fix the positions of all but cards 25, 26,
27 and 29 by the time statement 32 has been made. Note that statement
34 is about cards 25 and 26 rather than cards 25 and 27 as the pattern
may suggest. The declaration that there are 0 cards between cards 25 and
26 places card 29 next to card 30 and complete the determination of the
relative positions of the 52 cards. Thus 34 statements are indeed sufficient.

We conclude this section with another classic result. There is a unique
tree with 2 vertices, namely a 1-path, and a unique tree with 3 vertices,
namely, a 2-path. There are two trees with 4 vertices, a 3-star and a 3-path.
They are three trees with 5 vertices, a 4-star, a 4-path and the one shown
in Figure 7.10.

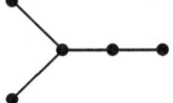

Figure 7.10

We now label the vertices with consecutive integers starting from 1. For the 1-path, there is a unique labeling by symmetry. For the 2-path, there are 3 different labelings, according to the label of the middle vertex. For the 3-star or the 4-star, there are 4 or 5 different labelings respectively, according to the label of the central vertex. For the 3-path, the central two vertices can be labeled in 6 ways and the two outer vertices in 2 ways afterwards. For the 4-path, the central vertex can be labeled in 5 ways, the two vertices adjacent to it in 6 ways and the two outer vertices in 2 ways. For the tree in Figure 7.10, the vertex with degree 3 can be labeled in 5 ways, the vertex with degree 2 in 4 ways and the vertex adjacent to it in 3 ways.

In summary, the number of labeled trees are 1 for those with two vertex, 3 for those with three vertices, $4 + 6 \times 2 = 16$ for those with four vertices, and $5 + 5 \times 6 \times 2 + 5 \times 4 \times 3 = 125$ for those with five vertices. These numbers follow a definite pattern.

Cayley's Formula for Labeled Trees.
The number of labeled trees with n vertices is n^{n-2}.

Proof:
The key idea is that n^{n-2} counts the number of sequences of length $n - 2$ in which each term is one of 1, 2, ..., n, with repetitions allowed. We wish to establish a one-to-one correspondence between such sequences and the labeled trees. We first give an algorithm which transform the labeled trees into sequences. By the Corollary to the Tree Formula, every tree has at least two terminal vertices. Choose the one with the largest label and delete it along with the lone edge incident with it. Record the label of the other vertex with which the deleted edge is incident. The resulting subgraph is still a tree, and again has at least two terminal vertices. Continue until a single edge remains. We will then have a sequence of length $n - 2$.

We illustrate this with three examples, as shown in Figure 7.11. For the labeled 5-star, the vertices removed are labeled 6, 4, 3 and 2 in that order, generating the sequence 5555. For the second example, the vertices removed are labeled 6, 3, 4 and 2, generating the sequence 4455. For the third example, the vertices removed are labeled 5, 3, 4 and 2 in that order, generating the sequence 3466.

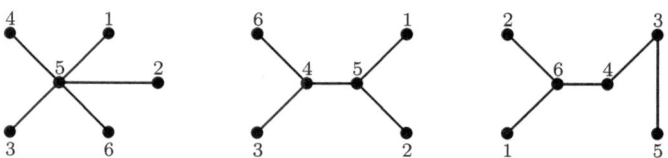

Figure 7.11

We now give the reverse algorithm which transform the sequences into labeled trees. For each sequence, a vertex of the desired labeled tree is said to be exposed if its label does not appear in the sequence. Delete the exposed vertex with the largest label, and attach it to the vertex whose label is the first term in the sequence. Then delete this label from the sequence, and if this is the only appearance of the label in the sequence, the corresponding vertex now becomes exposed. Continue until the sequence is empty. The two vertices not deleted are joined by the last edge of the labeled tree.

Labels in Sequence	Exposed Vertices	Deleted Vertices	Constructed Edges
1452	3,6	6	61
452	1,3	3	34
52	1,4	4	45
2	1,5	5	52
4244	1,3,5,6	6	64
244	1,3,5	5	52
44	1,2,3	3	34
4	1,2	2	24
4626	1,3,5	5	54
626	1,3,4	4	46
26	1,3	3	32
6	1,2	2	26

We illustrate with three examples, the sequences being 1452, 4244 and 4626 respectively. The transformations are recorded in the chart above. The resulting labeled trees, the first of which is a labeled 5-path, are shown in Figure 7.12.

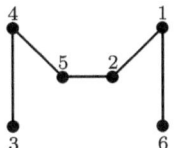 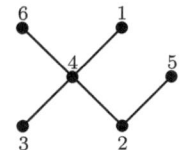 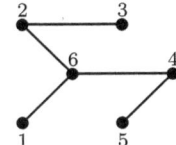

Figure 7.12

Exercises

4. Every square on a piece of graph paper divided into squares is painted in one of 23 colors, each used at least once. Two colors are friendly if there are two squares sharing a common side which are painted in these two colors. What is the minimum number of friendly pairs of colors?

5. In a country, there are 1988 towns and 4000 roads, each road connecting 2 towns. Prove that it is possible to making a tour of at most 20 towns by following the roads and returning to the starting town.

6. Each of the letters A, B, C, D, E, F and G is to be represented by a sequence of dots and dashes. Each dot takes half a second to transmit while each dash takes one second to transmit. In a 100-letter message, the seven letters appear 5, 6, 15, 16, 18, 19 and 21 times respectively. Design a code with no ambiguity such that the message may be transmitted in no more than 200 seconds.

Section 3. Directed Graphs

In a graph, an edge represents a symmetric relation between the two entities represented by the vertices with which the edge is incident. When this relation is not symmetric, such as one-way traffic between two junctions or the result of a game between two players, we need to impose an orientation on the edge.

A **directed** graph or **di-graph** is one in which the edges are directed. It goes from a **start** vertex to an **end** vertex. It contributes 1 to the **out-degree** of its start vertex, and 1 to the **in-degree** of its end vertex. A di-graph is connected if for any two vertices, there exists a directed path with the first as the initial vertex and the second as the final vertex.

Suppose in a city, if traffic along the one segment between any two junctions is closed, then it is still possible to get from anywhere to anywhere else in the city. It will be possible to convert all segments into one-way streets so that it is possible to get from anywhere to anywhere else.

The hypothesis is clearly necessary. If the closure of one segment disconnects the city into two mutually inaccessible parts, then it is only possible to go from one part to the other when this segment is converted into a one-way street. That the hypothesis is also sufficient is not immediately evident.

Robbin's Theorem.
If a graph remains connected when any one of its edges is deleted, then it can be converted into a connected di-graph.

Proof:
Let G be the graph and let v be any of its vertices. Consider the set of all subgraphs containing v for which the conversion is possible. This set is non-empty since it contains the subgraph consisting only of v. Now let H be the subgraph in this set containing the largest number of vertices. We may assume that H contains all the edges of G which are incident only with vertices in H. We claim that H is identical to G. Suppose this is not the case. Since G is connected, there exists a vertex u not in H such that it is joined to some vertex q in H. Since G is still connected when the edge qu is removed, there exists a path which connects u to some vertex p in H, possibly q again. We may assume that the other vertices on this path are not in H. Thus we can obtain a directed path from p to q via u. If we add this directed path to H, we obtain a connected di-graph containing v and having more vertices than H. This contradicts the maximality assumption on H.

A **tournament** is a complete di-graph. The vertices represent team participating in a round-robin tournament without draws, and the directed edges record the result of the games. It is somewhat surprising that every tournament has a directed Hamiltonian path.

Consider the longest directed path P in a tournament. Let the vertices on it be v_1, v_2, ..., v_n in that order. Suppose there is a vertex u not on P. If the edge between u and v_1 is directed from u, we may add u to the front of the path, which contradicts the maximality assumption on P. Similarly, the edge between u and v_n must be directed from u. Hence there exists a smallest index $k > 1$ such that the edge between u and v_k is directed from u. We can then extending the path by inserting u between v_{k-1} and v_k, which is a contradiction.

Problem 12.
Prove that a connected tournament with $n \geq 3$ vertices has a directed cycle of length k for every k, $3 \leq k \leq n$.

Solution:
We use mathematical induction on k. Let G be a strongly connected tournament. Pick an arbitrary vertex v. The remaining vertices may be divided into two sets X and Y, where vertices in X have edges going to v and vertices in Y have edges coming from v. If all edges between X and Y go from X to Y, then G is not strongly connected. Hence there is an edge going from some y in Y to some x in X. Then (v, y, x, v) is a directed cycle of length 3. Suppose we have a directed cycle of length k where $3 \leq k < n$. The remaining vertices may be divided into three sets X, Y and Z, where vertices in X have no edges coming from the cycle, vertices in Y have no edges going to the cycle and vertices in Z have both. We consider two cases.
Case 1. Z is empty.
Then both X and Y must be non-empty as otherwise G will not be connected. As in the basis, there is an edge going from some y in Y to some x in X. We can replace any vertex on the cycle with y and x and lengthen it by 1.
Case 2. Z is non-empty.
Then there exists some z in Z. Consider any edge leading to z from some vertex x on the cycle. Let y be the next vertex on the cycle. If we have an edge going from z to y, we can squeeze z into the cycle between x and y. Otherwise, consider the next vertex on the cycle. Eventually, we will find a place for z to be incorporated into the cycle.

Problem 13.
For any two members A and B in the Mathematical Sciences Society, either A owes B money or B owes A money. If A owes money to every other member, A is called the deadbeat. If for every member C, either B owes C money or B owes D money for some member D who owes C money, B is called a bum provided that B is not the deadbeat. Prove that the Society has either one deadbeat or at least three bums.

Solution:
For each member M, let X_M denote the set of members who owes M money, and Y_M those to whom M owes money. Let A be the member such that $|Y_A|$ is maximum. If X_A is empty, then A is the deadbeat. Suppose X_A is non-empty. We claim that A is a bum, that is, to every member of X_A, some member of Y_A must owe money. If some M in X_A owes money to everyone in Y_A, then $|Y_M| > |Y_A|$, a contradiction. Within X_A, there is a member B who is either the deadbeat or a bum. Since B owes A money directly and those in Y_A indirectly, B is a bum overall. If B is the deadbeat in X_A, the deadbeat or a bum within the set of members in Y_A who owe B money is a bum. If B is a bum within X_A, the deadbeat or a bum within the set of members in X_A who owe B money is a bum.

Problem 14.
The plan of a Martian underground is a closed self-intersecting curve, with at most one self-intersection at each point. Prove that a tunnel system for such a plan may be constructed in such a way that the train passes consecutively over and under the intersecting parts of the tunnel.

Solution:
We convert the underground into a di-graph. The vertices are the points of self-intersection, and the edges are directed from one vertex to another along the direction of the train. The edges of this di-graph divide the plane into non-overlapping regions. Each region represents either an urban area of a rural area. The urban areas are shaded in Figure 7.13 on the left. Since each vertex has in-degree 2 and out-degree 2, the designation of urban and rural area can be made, on a consistent basis, so that each edge separates an urban area from a rural area. We design the tunnel system according to the following rule. If the train approaches a junction with the rural area on its left, then it passes over the junction. Otherwise, it passes under. We now prove that this scheme works. Consider a typical situation as shown in Figure 7.13 on the right. Here the train approaches the junction J with the rural area on its right. Hence it will pass under. However, once across the junction, the rural area will now be on its left, so that it will pass over the next junction. Hence along the whole trip, it will be passing alternatively over and under junctions. Moreover, when the train returns to J, the rural area must be on its left, so that it will be passing over the junction. Hence we will not be running into difficulty at any junction.

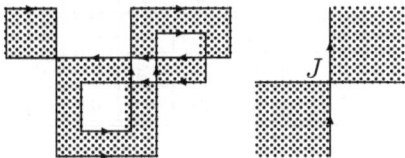

Figure 7.13

Problem 15. Ten friends send cards to one another, each sending cards to five different friends. Prove that at least two of them send cards to each other.

Solution:
Construct a di-graph with 10 vertices representing the 10 friends. A directed edge goes from vertex x to vertex y if the friend represented by x sends a card to the friend represented by y. Since each friend sends 5 cards to different friends, the total number of directed edges is $10 \times 5 = 50$. If no two friends send cards to each other, then the total number of directed edges is at most $\binom{10}{2} = 45$. This is a contradiction.

Problem 16.
In a town, each family lives in one apartment, and there are no vacant apartments. The Town Council wishes to move some or all of the families to other apartments. A family may only be moved once on any day, and only by exchanging apartments with another family. Prove that the Town Council's plan can be carried out in two days.

Solution:
Construct a di-graph with each vertex representing a family affected by the Town council's plan. An edge is drawn from vertex u to vertex v if the family represented by u is supposed to move to the apartment currently occupied by the family represented by v. Then the di-graph is a union of disjoint directed cycles, and we can handle each independently. We consider two cases.

Case 1. The length of the cycle is an even number $2n$.
On the first day, the families represented by v_k and v_{2n+1-k} exchange apartments. Then the families represented by v_n and v_{2n} are already in the correct apartments and can rest on the second day. Let $1 \le k \le n - 1$. The family represented by v_k moves to apartment $(2n+1) - k$, which is intended for the family represented by v_{2n-k}. On the other hand, the family represented by v_{2n-k} moves to apartment $(2n+1) - (2n-k) = k+1$, which is intended for the family represented by v_k. It follows that if the families represented by v_k and v_{2n-k} exchange apartments on the second day, then the Town council's plan has been carried out. The following chart illustrates the process for $n = 3$.

Apartment	1	2	3	4	5	6
Day 0	v_1	v_2	v_3	v_4	v_5	v_6
Day 1	v_6	v_5	v_4	v_3	v_2	v_1
Day 2	v_6	v_1	v_2	v_3	v_4	v_5

Case 2. The length of the cycle is an odd number $2n + 1$.

On the first day, the families represented by v_k and v_{2n+2-k} exchange apartments, with the family represented by v_{k+1} resting. Then the family represented by v_{2n+1} is already in the correct apartment and can rest on the second day. Let $1 \leq k \leq n$. The family represented by v_k moves to apartment $(2n+2) - k$, which is intended for the family represented by v_{2n+1-k}. On the other hand, the family represented by v_{2n+1-k} moves to apartment $(2n+2) - (2n+1-k) = k+1$, which is intended for the family represented by v_k. It follows that if the families represented by v_k and v_{2n+1-k} exchange apartments on the second day, then the Town council's plan has been carried out. The following chart illustrates the process for $n = 3$.

Apartment	1	2	3	4	5	6	7
Day 0	v_1	v_2	v_3	v_4	v_5	v_6	v_7
Day 1	v_7	v_6	v_5	v_4	v_3	v_2	v_1
Day 2	v_7	v_1	v_2	v_3	v_4	v_5	v_6

Problem 17.

Smallville is populated by unmarried men and women, some of them being mutual acquaintances. The City's two Official Matchmakers are aware of all the mutual acquaintances. One of them claimed: "I can arrange it so that every brown haired man will marry a woman with whom he is mutually acquainted." The other claimed, "I can arrange it so that every blond haired woman will marry a man with whom she is mutually acquainted." An amateur mathematician overheard their conversation and said, "Then both arrangements can be made at the same time!" Is that right?

Solution:

Construct a di-graph as follows. One group of vertices $x_1, x_2, \ldots, x_n$ represent the brown haired men. Another group of vertices $y_1, y_2, \ldots, y_m$ represent the blond haired women. From each of these vertices, draw an edge pointing to the vertex representing the mate promised by the appropriate matchmaker. If a man represented by x_k is promised a woman who is not blond haired, a new vertex w_k is introduced to represent the mate. Similarly, if a woman represented by y_k is promised a man who is not brown haired, a new vertex z_k is introduced to represent the mate. Now some of these edges may form a cycle. All vertices on such a cycle are x-vertices or y-vertices, as w-vertices and z-vertices have no out-going edges. Moreover, since the directed edges point alternately at the two groups, the cycle must be of even length. Hence the amateur mathematician can prescribe marriages, according to alternate edges along the cycle, between brown haired men and blond haired women. There are also edges which form a path. Such a path must consist of x-vertices and y-vertices, except that it terminates at a w-vertex or an z-vertex.

The amateur mathematician can prescribe marriages according to alternate edges along the path, starting with the initial edge. If the path is of even length, all marriages here are between brown haired men and blond haired women. If the path is of odd length, this is still the case except for the marriage corresponding to the last edge. In any cases, all brown haired men and all blond haired women are married off with mutual acquaintances.

Problem 18.
Several boxes are placed along a circle. Each box may contain any number of chips, including zero. A move consists of taking all the chips from some box and placing them in the subsequent boxes clockwise, one chip in every box, beginning from the next box in the clockwise direction. In each move after the first one, one must take the chips from the box in which the last chip was placed on the previous move. Prove that after several moves, the initial distribution of the chips among the boxes will reappear.

Solution:
We construct a di-graph as follows. Each vertex represents a distribution of chips coupled with the box from which chips must be taken. An edge goes from vertex a to vertex b if the state represented from a leads to the state represented by b. The out-degree of each vertex is clearly 1. We claim that so is its in-degree. From any given state, we can reverse the process by taking a chip from the box which is the last one to receive a chip, collecting 1 chip from each subsequent box counterclockwise, and stopping when we reach an empty box. We can only deposit all the chips into this box, and mark it as the one from which chips must be taken. This is the unique state which leads to the given one, and the claim is justified. Since the number of vertices is finite, the di-graph is a union of disjoint directed cycles. It does not matter which cycle contains the vertex which represents the initial state, as all states represented by vertices in this cycle will reappear.

Problem 19.
Construct a ring of twenty-seven digits each of which is 0, 1 or 2, such that when we read off three digits in clockwise order, all twenty-seven triples are distinct.

Solution:
Figure 7.14 shows a di-graph with nine vertices representing the binary sequences 00, 01, 02, 10, 11, 12, 20, 21 and 22. Each desired triple is represented by an edge xyz which goes from the vertex xy to the vertex yz. Since each vertex has in-degree 3 and out-degree 3, the di-graph has an Eulerian cycle. If we record the initial digit of each edge along such a cycle, we will obtain a desired ring.

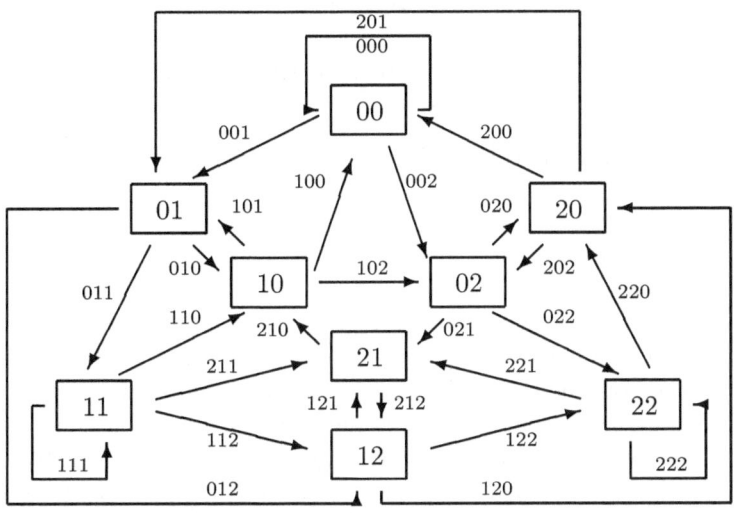

Figure 7.14

Exercises

7. Prove that in a round-robin tournament with no draws, there exist four teams A, B, C and D such that A has beaten B, C and D, B has beaten C and D, and C has beaten D.

8. A new website registered 2000 people. Each of them invited 1000 other registered people to be their friends. Two people are considered to be friends if and only if they have invited each other. What is the minimum number of pairs of friends on this website?

9. Several boxes are placed along a circle. Each box may contain any number of chips, including zero. A move consists of taking all the chips from any box and placing them in the subsequent boxes clockwise, one chip in every box, beginning from the next box in the clockwise direction. Is it true that for every initial distribution of the chips, one can get any possible distribution by performing an appropriate sequence of moves?

Bibliography

[1] Martin Gardner, The Sixth Book of Mathematical Diversions from Scientific American, University of Chicago Press, Chicago, 1971.

[2] Martin Gardner, The Last Recreations, Sprionger-Verlag, New York, 1997.

Solution to Exercises

1. Construct a graph with twenty vertices representing the teams. Two vertices are joined by a red edge if the two teams they represent play each other on the first day, and by a blue edge if they play on the second day. Since each vertex is incident with one red edge and one blue edge, each component of the graph is an even cycle. By taking every other vertex in each cycle, we have ten independent vertices. They represent ten teams no two of which have yet played each other.

2. If we use the standard checkerboard coloring extended to the third dimension in a $2 \times 2 \times 2$ cube, each unit cube is indeed sharing a common face with exactly three other unit cubes of opposite colors. We can use this as a building block to construct an $n \times n \times n$ cube when n is even. Just make sure that when two blocks share a common 2×2 face, all four unit cubes on one side share a common face with a unit cube of the same color on the other side. We now prove that the task is not possible if n is odd. Suppose to the contrary such an $n \times n \times n$ cube exists. Construct a graph with n^3 vertices representing the n^3 unit cubes. Two vertices are joined by an edge if they represent two unit cubes of opposite colors sharing a common face. Then every vertex is of degree 3. This violates the Corollary to the Parity Theorem.

3. Each of the four corners of the village is incident with two roads and requires at least one visit. Each of the remaining twelve intersections is incident with three or four sections of roads and requires at least two visits. Hence the minimum is at least $4 + 12 \times 2 = 28$. Figure 7.15 shows a closed route of length 28.

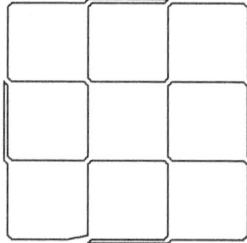

Figure 7.15

4. Construct a graph with 23 vertices representing the 23 colors. Two vertices are joined by an edge if the colors they represent are friendly. Since the piece of graph paper is connected, the constructed graph must also be connected. By the Connectivity Theroem, this graph has at least 22 edges. This minimum can be attained if we paint one square with each of 22 colors so that no two of these squares share a common side, and paint the remaining squares with the 23rd color.

5. Construct a graph G with 1988 vertices representing the towns, and 4000 edges representing the roads. An edge joins vertices x and y if a road connects the towns they represent. Delete all vertices of degree at most 2, and all edges incident with them. Continue until all remaining vertices are of degree at least 3. Since $4000 > 2 \times 1988$, the resulting graph is not empty. We construct a spanning tree of any connected component H by labeling any vertex 0. An unlabeled vertex adjacent to a vertex labeled k will be labeled $k+1$. Continue until all vertices of H have been labeled. Suppose no closed tour passing through at most 20 vertices exist in H. Then there is a unique path from the vertex labeled 0 to each vertex labeled k for $1 \leq k \leq 10$. For $k \geq 1$, there are at least $3 \times 2^{k-1}$ vertices labeled k. Hence the total number of vertices in H is at least $3(2^0 + 2^1 + 2^2 + \cdots + 2^9) = 3069 > 1988$. This is a contradiction.

6. We follow a binary tree with the upper branch representing dot and the lower branch representing dash. As soon as a code word has been chosen for a letter, that branch is sealed, which guarantee no ambiguity. The following code requires 195 seconds to transmit the message.

Letter	Frequency	Code Word	Unit time	Total Time
A	5	⊙ ⊖ ⊖	2.5	12.5
B	6	⊖ ⊙ ⊖	2.5	15
C	15	⊙ ⊙ ⊖	2	30
D	16	⊙ ⊖ ⊙	2	32
E	18	⊖ ⊙ ⊙	2	36
F	19	⊖ ⊖	2	38
G	21	⊙ ⊙ ⊙	1.5	31.5

7. Among the eight teams, there must be one which has won no fewer games than any other team. Call this team A. A must have won at least four games. Consider the mini-tournament among the four teams beaten by A. There must also be one which has won no fewer games than the other three. Call this team B. B must have won at least two games in this mini-tournament. The two teams B has beaten play each other. Call the winner C and the loser D. Then we have four teams with the desired properties.

8. Pretend that the 2000 people are seated at a round table, evenly spaced. Each invites the next 1000 people in clockwise order. Then only two people who are diametrically opposite to each other become friends. This shows that the number of pairs of friends may be as low as 1000. Construct a di-graph with 2000 vertices representing the people. Each vertex is incident to 1000 outgoing edges representing the invitations. The total number of edges is 2000×1000. The total number of pairs of vertices is $2000 \times 1999 \div 2 = 1999 \times 1000$. Even if every pair of vertices is connected by an edge, we still have $2000 \times 1000 - 1999 = 1000$ extra edges. These can only appear one of two double edges going in opposite directions. It follows that there must be at least 1000 reciprocal invitations, and therefore at least 1000 pairs of friends.

9. We construct a di-graph as follows. Each vertex represents a distribution of the chips. An edge goes from vertex A to vertex B if it is possible to change the distribution represented by A directly into the distribution represented by B. The out-degree and the in-degree of each vertex are both equal to the number of non-empty boxes in the distribution represented by that vertex. Hence the di-graph consists of connected components. It has only one component because the distribution in which all chips are in a specific box is reachable from any other distribution. We simply do not take chips from that box. Hence we can change any distribution into any other distribution.

Chapter Eight: Beanstalks

Section 1. Red and Blue Beanstalks

A **beanstalk** is a vertical column of segments with the bottom one resting on a horizontal ground. The game of **Red and Blue Beanstalks** is played between two players, Red and Blue respectively, on a finite collection of beanstalks where each segment is either red or blue. Turns alternate between the players. In each turn, the moving player removes a segment of her or his color, along with all segments above it. In other words, segments detached from the ground will disappear. The player who finds herself or himself without a move is the loser.

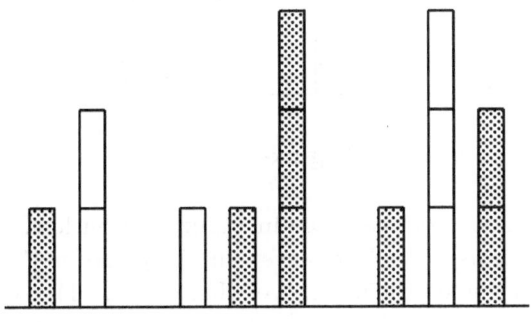

Figure 8.1

Figure 8.1 shows the opening positions of three games. The red segments are shaded while the blue segment are not. In the game on the left, if Red moves first, he must remove the only red segment and will lose in his next turn. If Blue moves first, she will be very stupid if she removes the bottom blue segment. By removing the top blue segment, she preserves another move for herself after Red's response, and wins. Henceforth, we assume that both players make intelligent moves.

A simple tactic emerges. If a player removes a segment from a beanstalk, she or he should remove the highest segment of her or his color. It is not hard to see that Red will win the game in the middle, regardless of who moves first. In the game on the right, whoever moves first is the loser.

In our analysis of the game, we take on a biased stance by always looking at things from Blue's viewpoint. A game has a positive value if she wins regardless of who moves first, and a negative value if she loses regardless of who moves first. There are also games in which whoever moves first loses, such as the empty game with no beanstalks, or games in which there is perfect symmetry between the red segments and the blue segments. Such neutral games are assigned the value 0.

© Springer International Publishing AG 2018
A. Liu, *S.M.A.R.T. Circle Minicourses*, Springer Texts in
Education, https://doi.org/10.1007/978-3-319-71743-2_8

There are also degrees of positiveness and negativeness. Red certainly have a bigger advantage over Blue in the second game than Blue has over Red in the first game. It is not hard to see that the value of the first game is $(-1)+2 = 1$ while that of the second game is $1+(-1)+(-3) = -3$. In fact, it would appear that the game value is just the difference when the total number of red segments subtracted from the total number of blue segments. This would be the case if all segments in each beanstalk are of the same color.

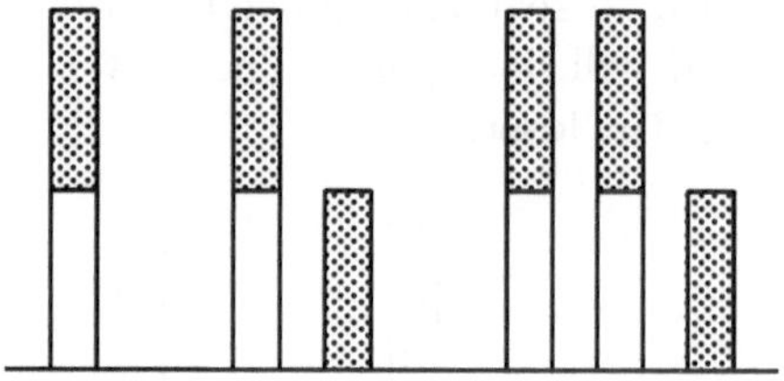

Figure 8.2

What is the value x of the game in Figure 8.2 on left? Clearly $x > 0$ since Blue is guaranteed the last move. In the game in Figure 8.2 in the middle, Red is guaranteed the last move. Hence $x + (-1) < 0$ so that $x < 1$. The natural choice seems to be $x = \frac{1}{2}$.

This can in fact be justified. Consider the game in Figure 8.2 on the right. If Blue moves first, she can remove one red segment along with one of her own. However, in his turn, Red removes the other jeopardized red segment which is sitting on top of a blue segment. Now we are left with a symmetric game which is worth 0, and Red will win. On the other hand, if Red moves first, his best move is to remove a jeopardized red segment. Blue will then remove the blue segment under the other red segment. The resulting game is once again symmetric, and Blue will win. It follows that $x + x + (-1) = 0$, so that indeed $x = \frac{1}{2}$.

We could have worked out the value x by the following rules:
(1) The value of a blue segment is positive.
(2) The value of a red segment is negative.
(3) A segment on the first or ground level has an absolute value of 1.
(4) A bean on the second level has an absolute value of $\frac{1}{2}$.
We will then have $x = 1 - \frac{1}{2} = \frac{1}{2}$.

The first three rules are clearly correct. The fourth is also correct under the assumption that the bottom two segments of a beanstalk must be of opposite colors. Otherwise, a beanstalk consisting of two blue segments would only be worth $1 + \frac{1}{2} = \frac{3}{2}$ instead of 2. If the bottom two segments of a beanstalk are of the same color, we can move the bottom one out and make it into a new beanstalk all by itself. Henceforth, we will assume that the bottom two segments are always of opposite color.

The third and fourth assumptions may be generalized to the following result.

Fundamental Theorem.

A bean on the nth level has an absolute value of $\frac{1}{2^{n-1}}$.

According to the Fundamental Theorem, the value of a beanstalk consisting of two red segments on top of a blue one is $x = 1 + (-\frac{1}{2}) + (-\frac{1}{4}) = \frac{1}{4}$. Let us verify this via the game in the Figure 8.3. We already know by symmetry that the value of a beanstalk consisting of a blue segment on top of a red one is $-\frac{1}{2}$.

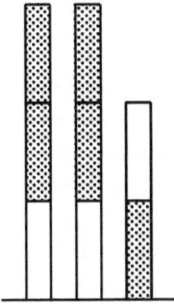

Figure 8.3

Suppose Red moves first. He will remove a red segment on the third level. Blue can leave behind a symmetric game by taking off the blue segment still with two red segments on top. Suppose Blue moves first. She will remove the blue segment on the second level. Red will remove a red segment on the third level. Blue will remove the blue segment still with two red segments on top. Red can now leave behind a symmetric game by taking off the red segment on the second level. It follows that $x + x + (-\frac{1}{2}) = 0$, so that indeed $x = \frac{1}{4}$.

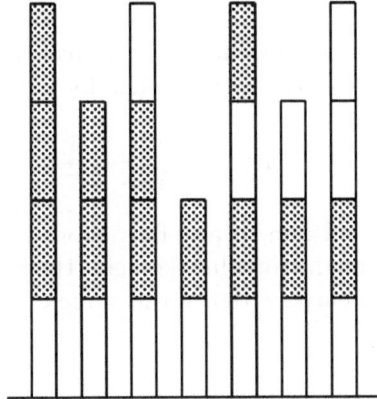

Figure 8.4

By playing appropriate games, the reader can verify that the beanstalks in Figure 8.4 have values $\frac{1}{8}$, $\frac{1}{4}$, $\frac{3}{8}$, $\frac{1}{2}$, $\frac{5}{8}$, $\frac{3}{4}$ and $\frac{7}{8}$ from left to right.

We now give the proof of the Fundamental Theorem.

Proof:
We use mathematical induction. We have already verified all the cases up to the second level. Let $n \geq 2$ be an integer. Suppose the absolute value of a segment on level k is $\frac{1}{2^{k-1}}$ for any positive integer $k \leq n$. We now prove that the value of a blue segment on level $n+1$ is $\frac{1}{2^n}$, regardless of the colors of the segments below it. Consider a beanstalk with a blue segment on top of n other segments. There are two cases.

Case 1. The bottom segment of the beanstalk is red.
Divide the beanstalk into an upper part consisting of the top red segment and everything above it, and a lower part of height $h \geq 0$, consisting of everything below the top red segment. We know the value y of the lower part from the induction hypothesis. It is either 0 or a negative fraction whose denominator is at most 2^n. Hence $1 - 2^n y$ is a positive integer. Let the value of the upper part be x. Note that $x < 0$. We claim that $x = -\frac{1}{2^n}$. Consider a game with 2^n copies of this beanstalk along with $1 - 2^n y$ beanstalks each consisting of a single blue segment. In the first 2^{n-1} turns, Blue will remove a blue segment on the $(n+1)$st level each time while Red will remove the red segment on the upper part of an untouched beanstalk each time, no matter who starts. We are left with a game with no segments on the $(n+1)$st level, so that the induction hypothesis can be applied. The $1 - 2^n y$ single blue segments have not been touched, and they have a combined value of $1 - 2^n y$. Each of the 2^{n-1} beanstalks from which Red has removed the red segment in the upper part is worth y. The upper part of each of the remaining 2^{n-1} beanstalks is worth

$$\frac{1}{2^{n-1}} + \frac{1}{2^{n-2}} + \cdots + \frac{1}{2^{h+1}} - \frac{1}{2^h} = -\frac{1}{2^{n-1}}.$$

Note that this value does not depend on h. Hence the whole beanstalk is worth $y - \frac{1}{2^{n-1}}$. The value of the game is therefore

$$1 - 2^n y + 2^{n-1} y + 2^{n-1}\left(y - \frac{1}{2^{n-1}}\right) = 0.$$

This means that the original game is also fair. From $1 - 2^n y + 2^n(y+x) = 0$, we have $x = -\frac{1}{2^n}$. Without the top blue segment, the value of the upper part is $-\frac{1}{2^{n-1}}$ as we have seen earlier. Hence the value of the top blue segment is $-\frac{1}{2^n} + \frac{1}{2^{n-1}} = \frac{1}{2^n}$.

Case 2. The bottom segment of the beanstalk is blue.

Divide the string into an upper part consisting of the top red segment and everything above it, and a lower part of height $h > 0$, consisting of everything below the top red segment. We know the value y of the lower part from the induction hypothesis. Let the value of the upper part be x. Note that $y > 0$ while $x < 0$. We claim that $x = -\frac{1}{2^n}$. Consider a game with 2^n copies of this beanstalk along with $2^n m - 1$ beanstalks each consisting of a single red segment. In the first 2^{n-1} turns, Blue will remove a blue segment on the $(n+1)$st level each time while Red will remove the red segment on the upper part of an untouched beanstalk each time, no matter who starts. We are left with a game with no segments on the $(n+1)$st level, so that the induction hypothesis can be applied. The $2^n y - 1$ single red segments have not been touched, and they have a combined value of $1 - 2^n y$. The remaining part of the argument is identical to that in Case 1, and the Fundamental Theorem has been proved.

Exercises

1. A beanstalk with at least three levels is such that the segment on the second level is different in color from the segments on the first and the third levels. Let x be its value. Determine in terms of x the value of the beanstalk obtained by removing the segment on the first level.

2. Prove that the game in Figure 8.5 is fair. This illustrates Case 1 of the Fundamental Theorem with $h = 0$.

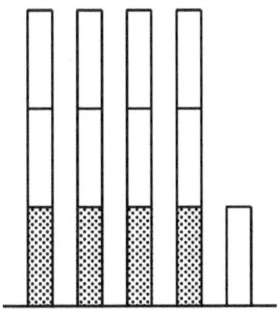

Figure 8.5

3. Prove that the game in Figure 8.6 is fair. This illustrates Case 2 of the Fundamental Theorem with $h = 1$.

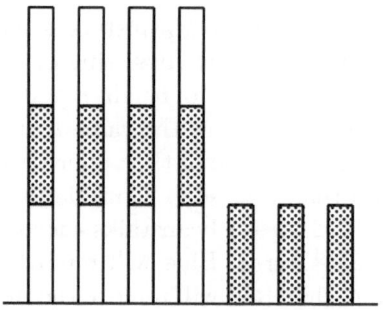

Figure 8.6

Section 2. Infinite Beanstalks

In this section, we consider beanstalk representations of real numbers. For rational numbers, we assume that the fractions are in the lowest terms.

In ordinary arithmetic, a real number is expressible as a terminating decimal if and only if it is a rational number whose denominator has only 2 or 5 as prime divisors.

In beanstalk arithmetic, it follows from the Fundamental Theorem that a real number is representable by a finite beanstalk if and only if it is a rational number whose denominator a power of 2.

In ordinary arithmetic, every rational number is expressible as an infinite decimal which is periodic. Finite decimals are also considered periodic with an empty period at the end.

In beanstalk arithmetic, every rational number is representable by an infinite beanstalk which is periodic. Finite beanstalks are also considered periodic with an empty period on top.

To establish this result, we need to come up with two algorithms, one to convert a rational number into a beanstalk, and the other to reverse the conversion. It is sufficient to focus on rational numbers between 0 and 1.

Consider as an example the rational number $\frac{1}{5}$, which is special in ordinary arithmetic but not in beanstalk arithmetic. We express both the numerator 1 and the denominator 5 as numbers in base 2, namely, 1 and 101, and perform the following long division.

$$
\begin{array}{r}
0.\ \ 0\ \ 0\ \ 1\ \ 1\ \ 0\ \ 0\ \ 1\ \ 1\ \ \cdots \\
101\ \ )1.\ \ 0\ \ 0\ \ 0\ \ 0\ \ 0\ \ 0\ \ 0\ \ 0\ \ \cdots \\
\hline
1\ \ 0\ \ 1 \\
\hline
1\ \ 1\ \ 0 \\
1\ \ 0\ \ 1 \\
\hline
1\ \ 0\ \ 0\ \ 0 \\
1\ \ 0\ \ 1 \\
\hline
1\ \ 1\ \ 0 \\
1\ \ 0\ \ 1 \\
\hline
1\ \ \cdots
\end{array}
$$

It follows that

$$
\frac{1}{5} = \qquad\qquad \frac{1}{8} + \frac{1}{16} \qquad\qquad + \frac{1}{128} + \frac{1}{256} + \cdots
$$

$$
= 1\ -\frac{1}{2}\ -\frac{1}{4}\ -\frac{1}{8}\ +\frac{1}{16}\ +\frac{1}{32}\ -\frac{1}{64}\ -\frac{1}{128}\ +\frac{1}{256} + \cdots.
$$

Figure 8.7 shows five copies of the beanstalk representation for $\frac{1}{5}$ along with a beanstalk consisting of a single red segment. This is a fair game.

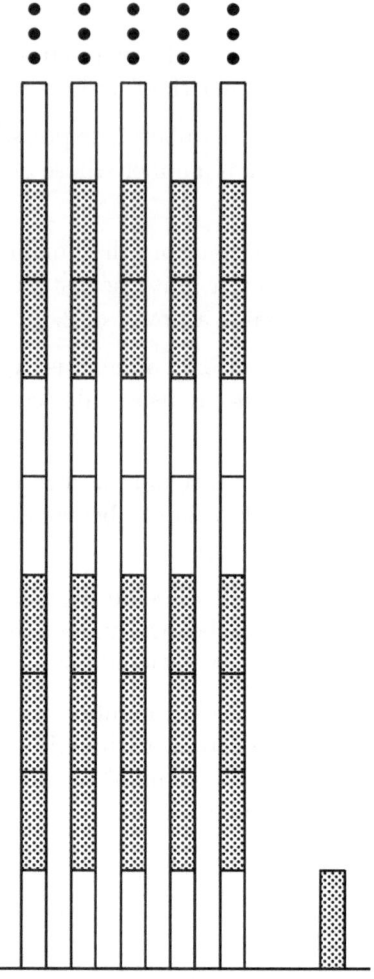

Figure 8.7

We now give the reverse algorithm which shows that the value of every infinite beanstalk which is periodic is a fraction whose denominator is not a power of 2. As an example, consider the infinite beanstalk with alternating blue and red segments, with a blue segment on the first level. Its value is

$$x = 1 - \frac{1}{2} + \frac{1}{4} - \frac{1}{8} + \cdots. \tag{1}$$

Note that $x = 1 - \left(\frac{1}{2} - \frac{1}{4}\right) - \left(\frac{1}{8} - \frac{1}{16}\right) - \cdots < 1$. On the other hand, $x = 1 - \frac{1}{2} + \left(\frac{1}{4} - \frac{1}{8}\right) + \cdots > 1 - \frac{1}{2}$. Similarly, we have $x < 1 - \frac{1}{2} + \frac{1}{4}$ and $x > 1 - \frac{1}{2} + \frac{1}{4} - \frac{1}{8}$, and so on.

To find the exact value of x, we may multiply both sides of (1) by 2 to obtain

$$2x = 2 - 1 + \frac{1}{2} - \frac{1}{4} + \frac{1}{8} - \cdots. \tag{2}$$

Adding (1) and (2) yields $3x = 2$, so that $x = \frac{2}{3}$.

The analysis of each algorithm involves summation to infinity, a process which uses the concept of limits in advanced mathematics. This is avoided within the context of the game of Red and Blue Beanstalks. To show that $\frac{2}{3}$ is the correct value for the above beanstalk, we only need to prove that the game in Figure 8.8 is fair. Despite the infiniteness of the beanstalks, this game is finite.

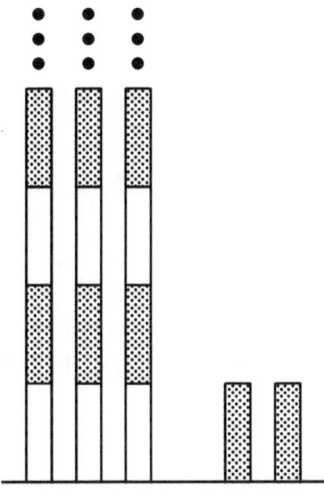

Figure 8.8

Suppose Blue starts. We may assume that she removes some blue segment from the first beanstalk, so that it now consists of $2k$ segments. Red can then remove a red segment from the second beanstalk so that it now consists of $2k + 1$ segments. By the Fundamental Theorem, the new value of the first beanstalk is $v_1 = 1 - \frac{1}{2} + \frac{1}{4} - \cdots + \frac{1}{2^{2k-1}} - \frac{1}{2^{2k}}$ while the new value of the second beanstalk is $v_2 = 1 - \frac{1}{2} + \frac{1}{4} - \cdots - \frac{1}{2^{2k}} + \frac{1}{2^{2k+1}}$.

Blue will lose because the value of the remaining game is

$$
\begin{aligned}
v_1 + v_2 + x - 2 &= 2 - 1 + \frac{1}{2} - \cdots + \frac{1}{2^{2k-2}} - \frac{1}{2^{2k-1}} + \frac{1}{2^{2k+1}} + x - 2 \\
&= x - \left(1 - \frac{1}{2} + \cdots - \frac{1}{2^{2k-2}} + \frac{1}{2^{2k-1}} - \frac{1}{2^{2k+1}}\right) \\
&= x - \left(1 - \frac{1}{2} + \cdots - \frac{1}{2^{2k-2}} + \frac{1}{2^{2k-1}} - \frac{1}{2^{2k}} + \frac{1}{2^{2k+1}}\right) \\
&< 0.
\end{aligned}
$$

Suppose Red starts. He will not touch the two red segments on the first level. We may assume that he removes some red segment from the first beanstalk, so that it now consists of $2k - 1$ segments. Blue can then remove a blue segment from the second beanstalk so that it now consists of $2k$ segments. Using a process analogous to the above, we can show that the value of the remaining game is positive, so that Red loses. It follows that the original game is fair, so that $3x - 2 = 0$ and $x = \frac{2}{3}$.

For irrational numbers, the decimal expressions are infinite and non-periodic. The beanstalk representations are also infinite and non-periodic. Just as we cannot write down the complete decimal, we cannot draw the entire beanstalks either. Suppose we agree to go as high as the sixth level. This representation can be obtained via rational approximation. We just illustrate with an example.

We have $0 < \frac{1}{4} < \frac{1}{2} < \frac{5}{8} < \frac{11}{16} < \frac{21}{32} < \frac{1}{\sqrt{2}} < \frac{23}{32} < \frac{13}{16} < \frac{3}{4} < \frac{7}{8} < 1$. Hence

$$
\begin{aligned}
\frac{1}{\sqrt{2}} &= \frac{1}{2} + \frac{1}{8} + \frac{1}{16} + \frac{8\sqrt{2} - 11}{16} \\
&= 1 - \frac{1}{2} + \frac{1}{4} - \frac{1}{8} + \frac{1}{16} + \frac{1}{32} - \cdots.
\end{aligned}
$$

In the truncated beanstalk, the segments on the first, third, fifth and sixth levels are blue while those on the second and the fourth level are red.

Exercises

4. Perform the long division in base 2 which leads to the beanstalk represenation for $\frac{2}{3}$

5. Verify that the game involving the beanstalk representation for $\frac{1}{5}$ is indeed a fair game.

6. Find the beanstalk representation for $\sqrt[3]{2}$ up to the sixth level.

Section 3. Beansprouts

The concepts of paths and trees have been defined in Chapter Seven. Whereas the underlying graph of a beanstalk is a path, a **beansprout** is a set of red and blue segments whose underlying graph is a tree with a single edge connected to the ground. Note that beanstalks are also beansprouts.

The game of Red and Blue Beansprouts is played in the same way. Unlike in Section 1, we no longer assume that the bottom two segments of a beanstalk have different colors. Each of the segments below the first two consecutive segments of different colors has absolute value 1. By symmetry, we assume that the bottom segment is blue.

We will confine our attention to finite beansprouts. and focus on finding the values of beansprouts which are not beanstalks. Such beansprouts with three segments all have the same shape, and there are three different coloring schemes, as shown in Figure 8.9.

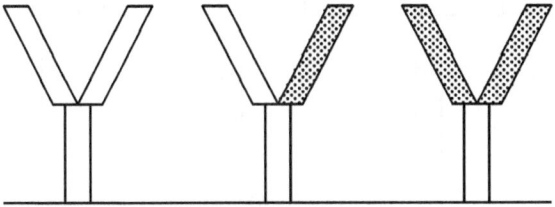

Figure 8.9

It is easy to see that these beansprouts have values 3, 1 and $\frac{1}{4}$ respectively. Suppose we have two or more segments with nothing on top and are resting on the same segment. The simple observation is that if two of them have different colors, they can be removed without affecting the value of the beansprout. If they are all of the same color, they can be stacked on top of one another.

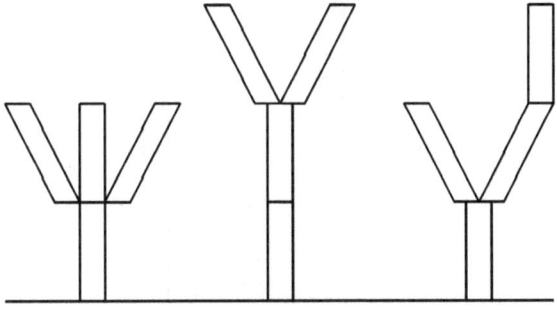

Figure 8.10

There are three different shapes of beansprouts with four segments. The two in Figure 8.10 on the left and in the middle can be handled using just the simple observation.

In the beansprout on the right, if three upper segments are all blue, the beansprout clearly has value 4. If they are all red, the value is $\frac{1}{8}$. Figure 8.11 shows the other six cases. Their respective values are $\frac{5}{2}$, 2, $\frac{3}{2}$, $\frac{3}{4}$, $\frac{1}{2}$ and $\frac{3}{8}$.

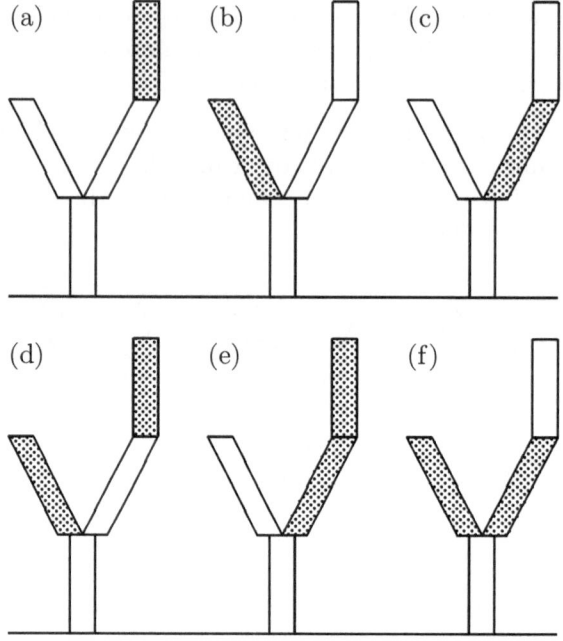

Figure 8.11

These can be verified by showing that certain games are fair. For case (a), we consider the game in Figure 8.12.

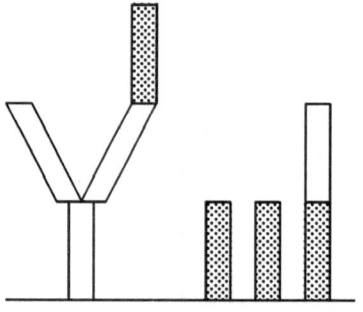

Figure 8.12

Regardless of who moves first, the blue segment sitting on top of a red segment and the red segment sitting on top of a blue segment will be removed in the first round. We are now left with three secured segments of each color. Whoever moves next will lose. Hence the game is fair, and the beansprout on the left indeed has value $\frac{5}{2}$.

However, we cannot continue working case by case. We need a general approach. The idea of the proof of the Fundamental Theorem in Section 1 is a reduction process by removing the top segment of a beanstalk. However, a beansprout may have several segments without nothing above. Nevertheless, by definition, it does have only one segment connected to the ground. Thus our reduction process removes this segment. Eventually the beansprout will decompose into a set of beanstalks. In Figure 8.13, we perform this reduction process on the seven trees with five edges and only one connected to the ground.

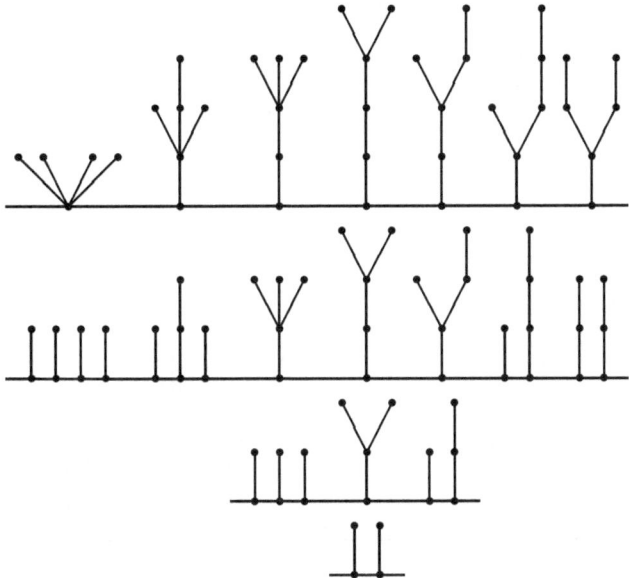

Figure 8.13

Suppose we have a beansprout of value x. We wish to determine the value of a new beansprout obtained by putting it on top of a segment connected to the ground. The proof of the following result, however, is beyond the scope of this book.

Basic Formula.

(1) If a beansprout of value x is put on top of a blue segment connected to the ground, the value of the new beansprout is $\frac{x+k}{2^k-1}$ where k is the smallest positive integer for which the numerator is greater than 1.

(2) If a beansprout of value x is put on top of a red segment connected to the ground, the value of the new beansprout is $\frac{x-k}{2^k-1}$ where k is the smallest positive integer for which the numerator is less than -1.

We now apply the reduction process and use the Basic Formula to determine the value of beansprouts.

For case (f) above, two beanstalks of respective values -1 and $-\frac{1}{2}$ sit on top of the bottom blue segment. From $-\frac{3}{2}+k>1$, we have $k=3$ so that the value of the beansprout is $\dfrac{-\frac{3}{2}+3}{4}=\dfrac{3}{8}$.

Consider now a slightly more complicated case such as the beansprout in Figure 8.14 on the left.

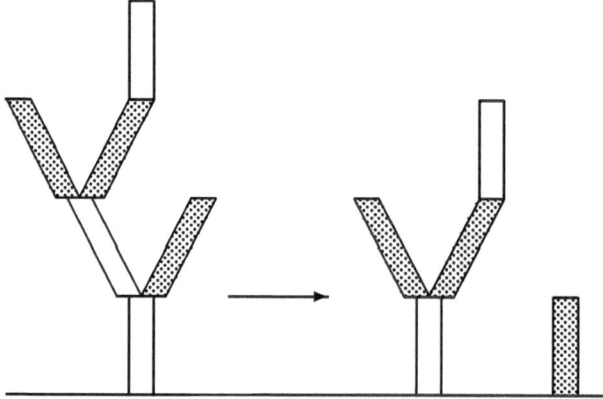

Figure 8.14

When the bottom blue segment is removed, we obtain the beansprout in Figure 8.14 on the right, along with a beanstalk consisting of a single red segment. When the bottom blue segment from this beansprout is removed, we obtain two beanstalks, one consisting of a single red segment, and one with a blue segment on top of a red segment.

The last two beanstalks have values -1 and $-\frac{1}{2}$ respectively. Hence what sits on top of the second removed blue segment has combined value $-\frac{3}{2}$. Since we require $-\frac{3}{2} + k > 1$, the smallest value is $k = 3$ and the beansprout has value $\dfrac{-\frac{3}{2} + 3}{2^2} = \dfrac{3}{8}$. Combining this with the single red segment which has value -1, what sits on top of the first removed blue segment has value $-\frac{5}{8}$. From $-\frac{5}{8} + k > 1$, we have $k = 2$ and the original beanstalk has value $\dfrac{-\frac{5}{8} + 2}{2} = \dfrac{11}{16}$.

We have generalized beanstalks to beansprouts. Further generalizations are wide open. For instance, if we replace the tree in the underlying structure of a beansprout with a general graph, we have a **beanbush**. The corresponding game is well-known. under the name **Hackenbush**. See [1], [2] and [3] for this and other generalizations, as well as the proof of the Basic Formula.

Exercises

7. Verify the values of the third beansprout with four segments for the cases (c) and (e) by proving that certain games are fair.

8. Use the Basic Formula to compute the values of the third beansprout with four segments for the cases (b) and (d).

9. Prove that the game in Figure 8.16 is fair.

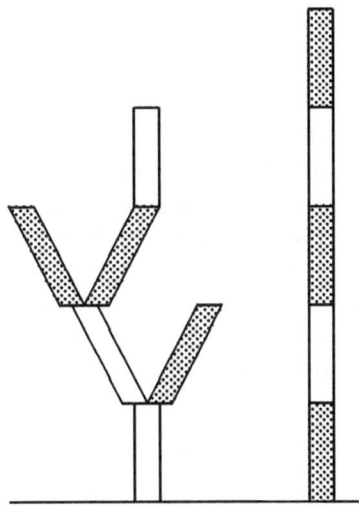

Figure

Bibliography

[1] E. R. Berlekamp, J. H. Conway and R. K. Guy, Winning Ways, A K Peters, Natick, 2001 to 2004.

[2] J. H. Conway, On Numbers and Games, A K Peters, Natick, 2001.

[3] Martin Gardner, Wheels, Life, and other Mathematical Amusements, W. H. Freeman, New York, 1983.

Solution to Exercises

1. Let y be the desired value. If the removed segment is blue, then $x = 1 + \frac{y}{2}$ so that $y = 2(x - 1)$. If the removed segment is red, then $x = -1 + \frac{y}{2}$ so that $y = 2(x + 1)$.

2. Suppose Blue starts. She removes a blue segment on the third level. Red removes a red segment that still has two blue segments on top of it. These two moves are repeated, and there are no segments left on the third level. The value of resulting game is $2(-\frac{1}{2}) + 1 = 0$. Since it is Blue's move, she will lose. Suppose Red starts. He removes any of the red segments. Blue removes a blue segment on the third level. Red removes a red segment that still has two blue segments on top of it, and Blue removes the remaining blue segment on the third level. We are then left with the same game as before, with value 0. Hence Red will lose, so that the game is fair.

3. Suppose Blue starts. She removes any of the blue segments on the third level. Red removes a red segment that still has a blue segment on top of it. These two moves are repeated, and there are no segments left on the third level. The value of resulting game is $2(-1-\frac{1}{2})+3 = 0$. Since it is Blue's move, she will lose. Suppose Red starts. He removes a red segment on the second level. Blue removes a blue segment on the third level. Red then removes a red segment that still has a blue segment on top of it, and Blue removes the remaining blue segment on the third level. We are then left with the same game as before, with value 0. Hence Red will lose, so that the game is fair.

4. We express both the numerator 2 and the denominator 3 as numbers in base 2, namely, 10 and 11, and perform the following long division.

$$
\begin{array}{r}
0.\ 1\ 0\ 1\ 0\ 1\ 0\ 1\ 0\ \cdots \\
\hline
11\ \overline{)10.\ 0\ 0\ 0\ 0\ 0\ 0\ 0\ 0\ \cdots} \\
1.\ 1 \\
\hline
1\ 0\ 0 \\
1\ 1 \\
\hline
1\ 0\ 0 \\
1\ 1 \\
\hline
1\ 0\ 0 \\
1\ 1 \\
\hline
1\ 0\ \cdots
\end{array}
$$

5. Let x be the value of each of the five identical beanstalks. Then

$$
x = 1 - \frac{1}{2} - \frac{1}{4} - \frac{1}{8} + \frac{1}{16} + \frac{1}{32} - \frac{1}{64} - \frac{1}{128} + \frac{1}{256} + \cdots
$$

and

$$
\begin{aligned}
3x &= 3 - \frac{3}{2} - \frac{3}{4} - \frac{3}{8} + \frac{3}{16} + \frac{3}{32} - \frac{3}{64} - \frac{3}{128} + \frac{3}{256} + \cdots \\
&= 2 - 1 - \frac{1}{2} - \frac{1}{4} + \frac{1}{8} + \frac{1}{16} - \frac{1}{32} - \frac{1}{64} + \frac{1}{128} + \cdots \\
&\quad + 1 - \frac{1}{2} - \frac{1}{4} - \frac{1}{8} + \frac{1}{16} + \frac{1}{32} - \frac{1}{64} - \frac{1}{128} + \frac{1}{256} + \cdots \\
&= 1 - \frac{1}{2} + \frac{1}{8} - \frac{1}{32} + \frac{1}{128} - \cdots \\
&= \frac{1}{2} + \frac{1}{4} - \frac{1}{8} - \frac{1}{16} + \frac{1}{32} + \frac{1}{64} - \frac{1}{128} - \cdots .
\end{aligned}
$$

Suppose Blue starts. We may assume that she removes from the first beanstalk some blue segment with a red segment directly above, but not the blue segment on the first level. It now consists of $4k + 1$ segments. Red can then remove a red segment from the second beanstalk so that it now consists of $4k + 3$ segments. By the Fundamental Theorem, the new value of the first beanstalk is

$$
v_1 = 1 - \frac{1}{2} - \frac{1}{4} - \frac{1}{8} + \frac{1}{16} + \cdots + \frac{1}{2^{4k-3}} - \frac{1}{2^{4k-2}} - \frac{1}{2^{4k-1}} + \frac{1}{2^{4k}}
$$

while the new value of the second beanstalk is

$$
v_2 = 1 - \frac{1}{2} - \frac{1}{4} - \cdots - \frac{1}{2^{4k-1}} + \frac{1}{2^{4k}} + \frac{1}{2^{4k+1}} - \frac{1}{2^{4k+2}} .
$$

Hence we have

$$
\begin{aligned}
& v_1 + v_2 - 1 \\
&= 2 - 1 - \frac{1}{2} - \frac{1}{4} + \frac{1}{8} + \cdots + \frac{1}{2^{4k-4}} - \frac{1}{2^{4k-3}} - \frac{1}{2^{4k-2}} + \frac{1}{2^{4k-1}} \\
& \quad + \frac{1}{2^{4k+1}} - \frac{1}{2^{4k+2}} - 1 \\
&= -\frac{1}{2} - \frac{1}{4} + \frac{1}{8} + \cdots + \frac{1}{2^{4k-4}} - \frac{1}{2^{4k-3}} - \frac{1}{2^{4k-2}} + \frac{1}{2^{4k-1}} \\
& \quad + \frac{1}{2^{4k}} - \frac{1}{2^{4k+1}} - \frac{1}{2^{4k+2}}
\end{aligned}
$$

Blue will lose because the value of the remaining game is

$$
v_1 + v_2 + 3x - 1 = -\frac{1}{2^{4k+3}} - \frac{1}{2^{4k+4}} + \cdots < 0.
$$

Suppose Red starts. He will not touch the red segments on the first level. We may assume that he removes from the first beanstalk a red segment with a blue segment directly above, so that it now consists of $4k - 1$ segments. Blue can then remove from the second beanstalk a blue segment with a red segment directly above, so that it now consists of $4k + 1$ segments. Using a process analogous to the above, we can show that the value of the remaining game is positive, so that Red loses. It follows that the original game is fair, so that $5x - 1 = 0$ and $x = \frac{1}{5}$.

6. We have $0 < \frac{1}{2} < \frac{5}{8} < \frac{3}{4} < \frac{25}{32} < \frac{11}{16} < \frac{1}{\sqrt[3]{2}} < \frac{13}{16} < \frac{27}{32} < \frac{7}{8} < 1$. Hence

$$
\begin{aligned}
\frac{1}{\sqrt{2}} &= \frac{1}{2} + \frac{1}{4} + \frac{1}{32} + \frac{16\sqrt[3]{4} - 25}{32} \\
&= 1 - \frac{1}{2} + \frac{1}{4} + \frac{1}{8} - \frac{1}{16} - \frac{1}{32} + \cdots.
\end{aligned}
$$

In the truncated beanstalk, the segments on the first, third and fourth levels are blue while those on the second, the fifth and the sixth level are red.

7. The Figure 8.17 is the game for case (c). If Blue moves first, she removes a blue segment sitting on top of a red segment. Then Red removes a red segment with a blue segment sitting on top of it. Each has two moves remaining, and Blue will lose. If Red moves first, he removes the red segment on top of a blue segment. Then Blue removes a blue segment sitting on top of a red segment. Each has two moves remaining, and Red will lose.

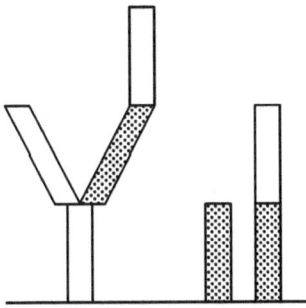

Figure 8.17

The Figure 8.18 is the game for case (e). Regardless of who moves first, the blue segment sitting on top of a red segment and the top red segment will be removed in the first round. Each has two moves remaining, and the next one to move will lose.

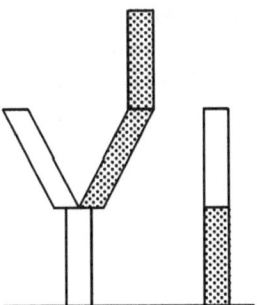

Figure 8.18

8. For case (b), two beanstalks of respective values -1 and 2 sit on top of the bottom blue segment. From $1 + k > 1$, we have $k = 1$ so that the value of the beansprout is $\dfrac{1 + 1}{1} = 2$. For case (d), two beanstalks of respective values -1 and $\frac{1}{2}$ sit on top of the bottom blue segment. From $-\frac{1}{2} + k > 1$, we have $k = 2$ so that the value of the beansprout is $\dfrac{-\frac{1}{2} + 2}{2} = \dfrac{3}{4}$.

9. Label the segments as shown in Figure 8.19. The red segments have odd labels and the blue segments have even labels.

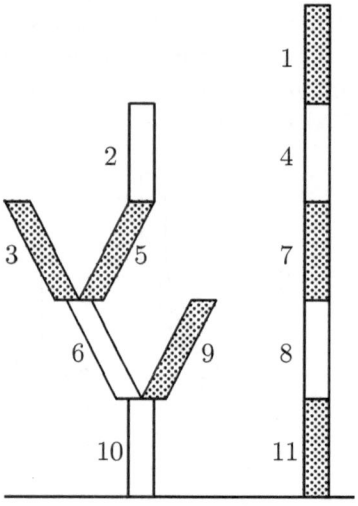

Figure 19

The beanstalk consisting of segments 1 and 4 neutralizes the beanstalk consisting of segments 2 and 5. Now Blue removes 6 if and only if Red removes 7, and Blue removes 8 if and only if Red removes 3. In the first case, the two remaining beanstalks neutralize each other. In the latter case, the segments 6 and 9 neutralizes each other, and the segments 10 and 11 neutralizes each other. Hence the game is indeed fair.

Chapter Nine: Inequalities

Section 1. The Rearrangement Inequality

We will introduce our subject via an example, taken from a Chinese competition in 1978.

"Ten people queue up before a tap to fill their buckets. Each bucket requires a different time to fill. In what order should the people queue up so as to minimize their combined waiting time?"

Common sense suggests that they queue up in ascending order of bucket-filling time. Let us see if our intuition leads us astray. We will denote by $T_1 < T_2 < \cdots < T_{10}$ the times required to fill the respective buckets.

If the people queue up in the order suggested, their combined waiting time will be given by $T = 10T_1 + 9T_2 + \cdots + T_{10}$. For a different queuing order, the combined waiting time will be $S = 10S_1 + 9S_2 + \cdots + S_{10}$, where $(S_1, S_2, \ldots, S_{10})$ is a permutation of $(T_1, T_2, \ldots, T_{10})$.

The two 10-tuples being different, there is a smallest index i for which $S_i \neq T_i$. Then $S_j = T_i < S_i$ for some $j > i$. Define $S_i' = S_j, S_j' = S_i$ and $S_k' = S_k$ for $k \neq i, j$. Let $S' = 10S_1' + 9S_2' + \cdots + S_{10}'$. Then

$$S - S' = (11 - i)(S_i - S_i') + (11 - j)(S_j - S_j') = (S_i - S_j)(j - i) > 0.$$

Hence the switching results in a lower combined waiting time.

If $(S_1', S_2', \ldots, S_{10}') \neq (T_1, T_2, \ldots, T_{10})$, this switching process can be repeated again. We will reach $(T_1, T_2, \ldots, T_{10})$ in at most 9 steps. Since the combined waiting time is reduced in each step, T is indeed the minimum combined waiting time.

We can generalize this example to the following result.

The Rearrangement Inequality.

Let $a_1 \leq a_2 \leq \cdots \leq a_n$ and $b_1 \leq b_2 \leq \cdots \leq b_n$ be real numbers. For any permutation $(a_1', a_2', \ldots, a_n')$ of $(a_1, a_2, \ldots, a_n)$, we have

$$a_1b_1 + a_2b_2 + \cdots + a_nb_n \geq a_1'b_1 + a_2'b_2 + \cdots + a_n'b_n$$
$$\geq a_nb_1 + a_{n-1}b_2 + \cdots + a_1b_n,$$

with equality if and only if $(a_1', a_2', \ldots, a_n')$ is equal to $(a_1, a_2, \ldots, a_n)$ or $(a_n, a_{n-1}, \ldots, a_1)$ respectively.

This can be proved by the switching process used in the introductory example. See for instance [2] or [6], which contain more general results. Note that unlike many inequalities, we do not require the numbers involved to be positive.

© Springer International Publishing AG 2018

A. Liu, *S.M.A.R.T. Circle Minicourses*, Springer Texts in Education, https://doi.org/10.1007/978-3-319-71743-2_9

Corollary 1.

Let $a_1, a_2, \ldots, a_n$ be real numbers and $(a_1', a_2', \ldots, a_n')$ be a permutation of $(a_1, a_2, \ldots, a_n)$. Then

$$a_1^2 + a_2^2 + \cdots + a_n^2 \geq a_1 a_1' + a_2 a_2' + \cdots + a_n a_n'.$$

Corollary 2.

Let $a_1, a_2, \ldots, a_n$ be positive numbers and $(a_1', a_2', \ldots, a_n')$ be a permutation of $(a_1, a_2, \ldots, a_n)$. Then

$$\frac{a_1'}{a_1} + \frac{a_2'}{a_2} + \cdots + \frac{a_n'}{a_n} \geq n.$$

Simple as it sounds, the Rearrangement Inequality is a result of fundamental importance. We shall derive from it many familiar and useful inequalities.

Example 1. The Arithmetic Mean Geometric Mean Inequality.

Let $x_1, x_2, \ldots, x_n$ be positive numbers. Then

$$\frac{x_1 + x_2 + \cdots + x_n}{n} \geq \sqrt[n]{x_1 x_2 \cdots x_n},$$

with equality if and only if $x_1 = x_2 = \cdots = x_n$.

Proof:

Let $G = \sqrt[n]{x_1 x_2 \cdots x_n}$, $a_1 = \frac{x_1}{G}$, $a_2 = \frac{x_1 x_2}{G^2}$, $\ldots$, $a_n = \frac{x_1 x_2 \cdots x_n}{G^n} = 1$. By Corollary 2,

$$n \leq \frac{a_1}{a_n} + \frac{a_2}{a_1} + \cdots + \frac{a_n}{a_{n-1}} = \frac{x_1}{G} + \frac{x_2}{G} + \cdots + \frac{x_n}{G},$$

which is equivalent to $\dfrac{x_1 + x_2 + \cdots + x_n}{n} \geq G$. Equality holds if and only if $a_1 = a_2 = \cdots = a_n$, or $x_1 = x_2 = \cdots = x_n$.

Example 2. The Geometric Mean Harmonic Mean Inequality.

Let $x_1, x_2, \ldots, x_n$ be positive numbers. Then

$$\sqrt[n]{x_1 x_2 \cdots x_n} \geq \frac{n}{\frac{1}{x_1} + \frac{1}{x_2} + \cdots + \frac{1}{x_n}},$$

with equality if and only if $x_1 = x_2 = \cdots = x_n$.

Proof:

Let $G, a_1, a_2, \ldots, a_n$ be as in Example 1. By Corollary 2,

$$n \leq \frac{a_1}{a_2} + \frac{a_2}{a_3} + \cdots + \frac{a_n}{a_1} = \frac{G}{x_1} + \frac{G}{x_2} + \cdots + \frac{G}{x_n},$$

which is equivalent to $G \geq \dfrac{n}{\frac{1}{x_1} + \frac{1}{x_2} + \cdots + \frac{1}{x_n}}$. Equality holds if and only if $x_1 = x_2 = \cdots = x_n$.

Example 3. The Root Mean Square Arithmetic Mean Inequality.
Let x_1, x_2, ..., x_n be real numbers. Then

$$\sqrt{\frac{x_1^2 + x_2^2 + \cdots + x_n^2}{n}} \geq \frac{x_1 + x_2 + \cdots + x_n}{n},$$

with equality if and only if $x_1 = x_2 = \cdots = x_n$.

Proof:
By Corollary 1, we have

$$x_1^2 + x_2^2 + \cdots + x_n^2 \geq x_1 x_2 + x_2 x_3 + \cdots + x_n x_1,$$
$$x_1^2 + x_2^2 + \cdots + x_n^2 \geq x_1 x_3 + x_2 x_4 + \cdots + x_n x_2,$$
$$\cdots \geq \cdots$$
$$x_1^2 + x_2^2 + \cdots + x_n^2 \geq x_1 x_n + x_2 x_1 + \cdots + x_n x_{n-1}.$$

Adding these and $x_1^2 + x_2^2 + \cdots + x_n^2 = x_1^2 + x_2^2 + \cdots + x_n^2$, we have

$$n(x_1^2 + x_2^2 + \cdots + x_n^2) \geq (x_1 + x_2 + \cdots + x_n)^2,$$

which is equivalent to the desired inequality. Equality holds if and only if $x_1 = x_2 = \cdots = x_n$.

Example 4. Cauchy's Inequality.
Let a_1, a_2, ..., a_n, b_1, b_2, ..., b_n be real numbers. Then

$$(a_1 b_1 + a_2 b_2 + \cdots + a_n b_n)^2 \leq (a_1^2 + a_2^2 + \cdots + a_n^2)(b_1^2 + b_2^2 + \cdots + b_n^2),$$

with equality if and only if for some constant $k, a_i = k b_i$ for $1 \leq i \leq n$ or $b_i = k a_i$ for $1 \leq i \leq n$.

Proof:
If $a_1 = a_2 = \cdots = a_n = 0$ or $b_1 = b_2 = \cdots = b_n = 0$, the result is trivial. Otherwise, define $S = \sqrt{a_1^2 + a_2^2 + \cdots + a_n^2}$ and $T = \sqrt{b_1^2 + b_2^2 + \cdots + b_n^2}$. Since both are non-zero, we may let $x_i = \frac{a_i}{S}$ and $x_{n+i} = \frac{b_i}{T}$ for $1 \leq i \leq n$. By Corollary 1,

$$
\begin{aligned}
2 &= \frac{a_1^2 + a_2^2 + \cdots + a_n^2}{S^2} + \frac{b_1^2 + b_2^2 + \cdots + b_n^2}{T^2} \\
&= x_1^2 + x_2^2 + \cdots + x_{2n}^2 \\
&\geq x_1 x_{n+1} + x_2 x_{n+2} + \cdots + x_n x_{2n} + x_{n+1} x_1 + x_{n+2} x_2 + \cdots + x_{2n} x_n \\
&= \frac{2(a_1 b_1 + a_2 b_2 + \cdots + a_n b_n)}{ST},
\end{aligned}
$$

which is equivalent to the desired inequality. Equality holds if and only if $x_i = x_{n+i}$ for $1 \leq i \leq n$, or $a_i T = b_i S$ for $1 \leq i \leq n$.

We now illustrate the power of the Rearrangement Inequality by giving simple solutions to a number of problems from the International Mathematical Olympiad.

Example 5. (1975)
Let $x_1 \leq x_2 \leq \cdots \leq x_n$ and $y_1 \leq y_2 \leq \cdots \leq y_n$ be real numbers. Let $(z_1, z_2, \cdots, z_n)$ be a permutation of $(y_1, y_2, \ldots, y_n)$. Prove that

$$(x_1 - y_1)^2 + (x_2 - y_2)^2 + \cdots + (x_n - y_n)^2 \leq (x_1 - z_1)^2 + (x_2 - z_2)^2 + \cdots + (x_n - z_n)^2.$$

Solution:
Note that we have $y_1^2 + y_2^2 + \cdots + y_n^2 = z_1^2 + z_2^2 + \cdots + z_n^2$. After expansion and simplification, the desired inequality is equivalent to

$$x_1 y_1 + x_2 y_2 + \cdots + x_n y_n \geq x_1 z_1 + x_2 z_2 + \cdots + x_n z_n,$$

which follows from the Rearrangement Inequality.

Example 6. (1978)
Let a_1, a_2, ..., a_n be distinct positive integers. Prove that

$$\frac{a_1}{1^2} + \frac{a_2}{2^2} + \cdots + \frac{a_n}{n^2} \geq \frac{1}{1} + \frac{1}{2} + \cdots + \frac{1}{n}.$$

Solution:
Rearrange $a_1, a_2, \ldots, a_n$ in non-descending order as $a_1' \leq a_2' \leq \cdots \leq a_n'$. Then $a_i' \geq i$ for $1 \leq i \leq n$. By the Rearrangement Inequality,

$$\begin{aligned}
\frac{a_1}{1^2} + \frac{a_2}{2^2} + \cdots + \frac{a_n}{n^2} &\geq \frac{a_1'}{1^2} + \frac{a_2'}{2^2} + \cdots + \frac{a_n'}{n^2} \\
&\geq \frac{1}{1^2} + \frac{2}{2^2} + \cdots + \frac{n}{n^2} \\
&\geq \frac{1}{1} + \frac{1}{2} + \cdots + \frac{1}{n}.
\end{aligned}$$

Example 7. (1964)
Let a, b and c be the sides of a triangle. Prove that

$$a^2(b + c - a) + b^2(c + a - b) + c^2(a + b - c) \leq 3abc.$$

First Solution:
We may assume that $a \geq b \geq c$. Then

$$\begin{aligned}
&3abc - a^2(b + c - a) - b^2(c + a - b) - c^2(a + b - c) \\
=\ &abc + a^3 - a^2b - a^2c + abc + b^3 - b^2c - b^2a \\
&+ abc + c^3 - c^2a - c^2b \\
=\ &a(b - a)(c - a) - b(c - b)(b - a) + c(c - a)(c - b) \\
=\ &a(b - a)(c - a) + (c - b)(c(c - a) - b(b - a)) \\
\geq\ &0.
\end{aligned}$$

Second Solution:
Let $a = y + z$, $b = z + x$ and $c = x + y$. Then the conditions on a, b and c imply that x, y and z are non-negative. By the Arithmetic-Geometric Mean Inequality, $abc \geq 2\sqrt{yz} \cdot 2\sqrt{zx} \cdot 2\sqrt{xy} = 8xyz$. Note that we have $b + c - a = 2x$, $c + a - b = 2y$ and $a + b - c = 2z$. Hence

$$
\begin{aligned}
abc &\geq (b + c - a)(c + a - b)(a + b - c) \\
&= -a^3 - b^3 - c^3 + a^2b + b^2c + c^2a + ab^2 + bc^2 + ca^2 - 2ab \\
&= a^2(b + c - a) + b^2(c + a - b) + c^2(a + b - c) - 2abc.
\end{aligned}
$$

The desired result follows immediately.

Third Solution:
We may assume that $a \geq b \geq c$. We first prove that

$$c(a + b - c) \geq b(c + a - b) \geq a(b + c - a).$$

Note that $c(a + b - c) - b(c + a - b) = (b - c)(b + c - a) \geq 0$. The other part of the inequality can be proved in the same manner. By the Rearrangement Inequality, we have

$$a^2(b{+}c{-}a){+}b^2(c{+}a{-}b){+}c^2(a{+}b{-}c) \leq ba(b{+}c{-}a){+}cb(c{+}a{-}b){+}ac(a{+}b{-}c),$$

$$a^2(b{+}c{-}a){+}b^2(c{+}a{-}b){+}c^2(a{+}b{-}c) \leq ca(b{+}c{-}a){+}ab(c{+}a{-}b){+}bc(a{+}b{-}c).$$

Adding these two inequalities, the right side simplifies to $6abc$. The desired inequality now follows.

Example 8. (1983)
Prove that $a^2b(a - b) + b^2c(b - c) + c^2a(c - a) \geq 0$, where a, b and c are the sides of a triangle.

First Solution:
We may assume that $a \geq \max b, c$. An expression that is non-negative is $a(b - c)^2(b + c - a)$. Now

$$
\begin{aligned}
&a^2b(a - b) + b^2c(b - c) + c^2a(c - a) - a(b - c)^2(b + c - a) \\
&= a^3b - a^2b^2 + b^3c - b^2c^2 + c^3a - c^2a^2 \\
&\quad -(ab^3 - a^2b^2 - ab^2c - abc^2 + 2a^2bc + ac^3 - a^2c^2) \\
&= a^3b - ab^3 - a^2bc + b^3c + abc^2 - b^2c^2 - a^2bc + ab^2c \\
&= b(a(a^2 - b^2) - c(a^2 - b^2) + c^2(a - b) - ca(a - b)) \\
&= b(a - b)(a^2 + ab - ac - bc + c^2 - ca) \\
&= b(a - b)((a - c)^2 + b(a - c)) \\
&= b(a - b)(a - c)(a + b - c) \\
&\geq 0.
\end{aligned}
$$

The desired result follows immediately.

Second Solution:

Let $a = y + z$, $b = z + x$ and $c = x + y$. Then the conditions on a, b and c imply that x, y and z are non-negative. Now

$$
\begin{aligned}
& a^2 b(a - b) + b^2 c(b - c) + c^2 a(c - a) \\
= \ & (y + z)^2 (z + x)(y - x) + (z + x)^2 (x + y)(z - y) \\
& + (x + y)^2 (y + z)(x - z) \\
= \ & (z^2 + yz + zx + xy)(y^2 + yz - zx - xy) \\
& + (x^2 + yz + zx + xy)(z^2 + zx - xy - yz) \\
& + (y^2 + yz + zx + xy)(x^2 + xy - yz - zx) \\
= \ & z^2(y^2 + yz - zx - xy) + x^2(z^2 + zx - xy - yz) \\
& + y^2(x^2 + xy + yz - zx) + (yz + zx + xy)(y^2 + yz - zx - xy \\
& + z^2 + zx - xy - yz + x^2 + xy - yz - zx) \\
= \ & z^2 y^2 + x^2 z^2 + y^2 x^2 + yz^3 + zx^3 + xy^3 - z^3 x - x^3 y - y^3 z \\
& - xyz^2 - x^2 yz - xy^2 z - y^2 z^2 - z^2 x^2 - x^2 y^2 - 2x^2 yz - 2xy^2 z - 2xyz^2 \\
& + x^2 yz + xy^2 z + xyz^2 + x^3 y + y^3 z + z^3 x + xy^3 + yz^3 + zx^3 \\
= \ & 2(xy^3 + yz^3 + zx^3) - 2xyz(x + y + z).
\end{aligned}
$$

By Cauchy's Inequality,

$$
\begin{aligned}
(z + x + y)(xy^3 + yz^3 + zx^3) & \geq (y\sqrt{xyz} + z\sqrt{xyz} + x\sqrt{xyz})^2 \\
& = xyz(y + z + x)^2.
\end{aligned}
$$

It follows that we have $xy^3 + yz^3 + zx^3 \geq xyz(x + y + z)$. This is equivalent to $a^2 b(a - b) + b^2 c(b - c) + c^2 a(c - a) \geq 0$.

Third Solution:

We may assume that $a \geq \max b, c$. As in the Third Solution of Example 7, we have $a(b + c - a) \leq b(c + a - b) \leq c(a + b - c)$ if $a \geq b \geq c$. By the Rearrangement Inequality,

$$
\begin{aligned}
& \frac{1}{c} a(b + c - a) + \frac{1}{a} b(c + a - b) + \frac{1}{b} c(a + b - c) \\
\leq \ & \frac{1}{a} a(b + c - a) + \frac{1}{b} b(c + a - b) + \frac{1}{c} c(a + b - c) \\
= \ & a + b + c.
\end{aligned}
$$

This simplifies to $\frac{1}{c} a(b - a) + \frac{1}{a} b(c - b) + \frac{1}{b} c(a - c) \leq 0$, which is equivalent to the desired inequality. Now $a(b + c - a) \leq c(a + b - c) \leq b(c + a - b)$ if $a \geq c \geq b$. All we have to do is interchange the second and the third terms of the second displayed line above.

Exercises

1. Let a_1, a_2, $\ldots$, a_n be positive numbers. Prove that

$$\frac{a_2}{a_1} + \frac{a_3}{a_2} + \cdots + \frac{a_n}{a_{n-1}} + \frac{a_1}{a_n} \geq n$$

   (a) using the Rearrangement Inequality;

   (b) without using the Rearrangement Inequality.

2. Let a_1, a_2, $\ldots$, a_n be any positive real numbers and let $(b_1, b_2, \ldots, b_n)$ be any permutation of $(a_1, a_2, \ldots, a_n)$. Prove that

$$\frac{a_1}{b_1} + \frac{a_2}{b_2} + \cdots + \frac{a_n}{b_n} \geq n$$

   (a) using the Rearrangement Inequality;

   (b) without using the Rearrangement Inequality.

3. For all positive numbers a, b and c, prove that $\frac{a}{b+c} + \frac{b}{c+a} + \frac{c}{a+b} \geq \frac{3}{2}$

   (a) using the Rearrangement Inequality;

   (b) without using the Rearrangement Inequality.

Section 2. The Majorization Inequality

This section is an elementary exposition on the important and useful concept of majorization. We will give some applications of the Majorization Inequality in problem-solving.

Let $S(n, s)$ denote the set of all sequences of n real numbers with sum s. We assume that the terms in each sequence are arranged in non-ascending order. We can define a partial ordering on $S(n, s)$ called majorization as follows.

Let $X = (x_1, x_2, \ldots, x_n)$ and $Y = (y_1, y_2, \ldots, y_n)$ be sequences in $S(n, s)$. We say that X majorizes Y if $x_1 + x_2 + \cdots + x_k \geq y_1 + y_2 + \cdots + y_k$ for $1 \leq k \leq n$. Symbolically, we write $X \geq Y$, and $X > Y$ unless $x_k = y_k$ for $1 \leq k \leq n$, in which case $X = Y$.

Example 9.
Verify that in $S(6, 23)$, $(7, 5, 5, 3, 2, 1) > (6, 6, 3, 3, 3, 2)$.

Solution:
We have $7 > 6$, $7+5 = 6+6$, $7+5+5 > 6+6+3$, $7+5+5+3 > 6+6+3+3$, $7+5+5+3+2 > 6+6+3+3+3$ and $7+5+5+3+2+1 = 23 = 6+6+3+3+3+2$.

Clearly, $X \geq X$ for all X in $S(n, s)$. Suppose $X \geq Y$ and $Y \geq Z$. Then

$$x_1 + x_2 + \cdots + x_k \geq y_1 + y_2 + \cdots + y_k \geq z_1 + z_2 + \cdots + z_k$$

for $1 \geq k \geq n$, so that $X \geq Z$. Suppose $X \geq Y$ and $Y \geq X$. Then

$$x_1 + x_2 + \cdots + x_k = y_1 + y_2 + \cdots + y_k$$

for $1 \geq k \geq n$, so that $X = Y$. Hence $\geq$ has the reflexive, transitive and anti-symmetric properties.

However, it is not a total ordering. For instance, neither $(8, 5, 5, 3, 1, 1)$ nor $(7, 7, 3, 3, 3, 0)$ in $S(6, 23)$ majorizes the other, as $8 > 7$ but $8+5 < 7+7$. On the other hand, every sequence in $S(n, s)$ majorizes $(\frac{s}{n}, \frac{s}{n}, \ldots, \frac{s}{n})$.

We now state and prove a simple but significant result in majorization.

The Fundamental Lemma.
Suppose $X > Y$ in $S(n, s)$.

(a) There exists Z in $S(n, s)$ such that $X > Z \geq Y$ and Z differs from X in exactly two terms.

(b) There exists a sequence $Z_0, Z_1, \ldots, Z_m$ in $S(n, s)$, $m < n$, such that $X = Z_0 > Z_1 > \cdots > Z_m = Y$, and Z_k differs from Z_{k-1} in exactly two terms for $1 \leq k \leq m$.

Proof:

Since $X > Y$, we must have $x_k > y_k$ for at least one k. Let i be the largest such value. Then we have $x_k \geq y_k$ for $i \leq k \leq n$. It follows from $x_1 + x_2 + \cdots + x_{i-1} \geq y_1 + y_2 + \cdots + y_{i-i}$ that $x_1 + x_2 + \cdots + x_i > y_1 + y_2 + \cdots + y_i$. Hence $x_k < y_k$ for some $k > i$. Let j be the smallest such value. Then we have $x_k = y_k$ for $i \leq k \leq j$. Let $d = \min\{x_i - y_i, y_j - x_j\}$. Then $d > 0$.

(a) Define Z by taking $z_i = x_i - d, z_j = x_j + d$ and $z_k = x_k$ for $k \neq i, j$. Clearly, Z is in $S(n, s)$. Moreover, $y_i < x_i - d = z_i$ and $y_i > x_i + d = z_i$. We claim that $z_1 \geq z_2 \geq \cdots \geq z_n$. The only doubtful cases involve z_i or z_j. If $i > 1$, we have $z_{i-1} = x_{i-1} \geq x_i > x_i - d = z_i$. If $j < n, z_{j+1} = x_{j+1} \leq x_i < x_i + d = z_j$. If $i < j - 1$, we have $z_i \geq y_i \geq y_{i+1} = x_{i+1} = z_{i+1}$ and $z_j \leq y_j \leq y_{j-1} = x_{j-1} = z_{j-1}$. Finally, if $i = j - 1, z_i \geq y_i \geq y_j \geq z_j$. Thus the claim is justified. For $1 \leq k < i$, we have $z_1 + z_2 + \cdots + z_k = x_1 + x_2 + \cdots + x_k > y_1 + y_2 + \cdots + y_k$. Since $x_i > z_i \geq y_i, x_1 + x_2 + \cdots + x_i \geq z_1 + z_2 + \cdots + z_i \geq y_1 + y_2 + \cdots + y_i$. For $i < k < j, z_k = x_k = y_k$, so that

$$x_1 + x_2 + \cdots + x_k > z_1 + z_2 + \cdots + z_k \geq y_1 + y_2 + \cdots + y_k.$$

Finally, since $z_i + z_j = x_i + x_j$, we have

$$z_1 + z_2 + \cdots + z_k = x_1 + x_2 + \cdots + x_k \geq y_1 + y_2 + \cdots + y_k$$

for $j \leq k \leq n$. It follows that $X > Z \geq Y$.

(b) Choose $Z_0 = X$ and $Z_1 = Z$ defined above. If $Z_1 = Y$, the task is completed. If $Z_1 > Y$, let Z_1 play the role of X and take the resulting Z as Z_2. This construction may have to be repeated a number of times. Since Z_k agrees with Y in at least one more term than Z_{k-1}, the task is finite, and it involves no more than $n - 1$ steps.

Example 10.

In $S(6, 23)$, let $X = (7, 5, 5, 3, 2, 1)$ and $Y = (6, 6, 3, 3, 3, 2)$. Find Z_0, Z_1, Z_2 and Z_3 such that $X = Z_0 > Z_1 > Z_2 > Z_3 = Y$ and Z_k differs from Z_{k-1} in exactly two terms for $1 \leq k \leq 3$.

Solution:

By Example 9, $X > Y$, so that the construction in the Fundamental Lemma can be applied. We take $Z_0 = X = (7, 5, 5, 3, 2, 1)$. Since $x_3 = 5 > 3 = y_3$ and $x_k \geq y_k$ for $3 < k \leq 6, i = 3$. Although $x_2 = 5 < 6 = y_2$, we do not take $j = 2$ because we must have $j > i$. Since $x_4 = 3 = y_4$ and $x_5 = 2 < 3 = y_5, j = 5$. Now $x_3 - y_3 = 2$ and $y_5 - x_5 = 1$. Hence $d = 1$ and we have $Z_1 = (7, 5, 4, 3, 3, 1)$. In the next step, $i = 3, j = 6, d = 1$ and $Z_2 = (7, 5, 3, 3, 3, 2)$. In the final step, $i = 1, j = 2, d = 1$ and we have $Z_3 = (6, 6, 3, 3, 3, 2) = Y$.

Before we can state the Majorization Inequality, we need another concept. A continuous real-valued function on one real variable **convex** over an interval if a chord joining any two points on the graph of the function within the interval lies on or above the graph itself. For most functions, it is sufficient to know that the midpoint of the chord lies on or above the graph. If all such midpoints lie above the graph, the function is said to be strictly convex.

That a function $f(x)$ is said to be convex may be expressed algebraically as follows. Let $(x_1, f(x_1))$ and $(x_2, f(x_2))$ be two distinct points on the graph of $f(x)$. The midpoint of the chord joining them has coordinates $(\frac{x_1+x_2}{2}, \frac{f(x_1)+f(x_2)}{2})$. The y-coordinate of the point on the graph having the same x-coordinate is $f(\frac{x_1+x_2}{2})$. Hence $f(x)$ is convex if and only if

$$\frac{f(x_1) + f(x_2)}{2} \geq f\left(\frac{x_1 + x_2}{2}\right)$$

for all x_1 and x_2 in its domain. Obviously, equality must hold if $x_1 = x_2$. If it does not hold in any other case, f is strictly convex.

Example 11.

Prove that $\frac{x}{2-x}$ is strictly convex on $(-\infty, 2)$.

Solution:

We have to prove that for all x_1 and x_2 in $(-\infty, 2)$,

$$\frac{\frac{x_1}{2-x_1} + \frac{x_2}{2-x_2}}{2} \geq \frac{\frac{x_1+x_2}{2}}{2 - \frac{x_1+x_2}{2}}.$$

Since $x_1 < 2$ and $x_2 < 2$, all of $2-x_1, 2-x_2$ and $4-x_1-x_2$ are positive. Now $x_1(2-x_2)(4-x_1-x_2) + x_2(2-x_1)(4-x_1-x_2) - 2(x_1+x_2)(2-x_1)(2-x_2)$ simplifies to $(x_1 - x_2)^2 \geq 0$, which is equivalent to the desired inequality. Equality holds if and only if $x_1 = x_2$.

Examples of other convex functions are x^2 over all real numbers, $\frac{1}{x}$ over the interval $(0, \infty)$, $\sec x$ over the interval $(-\frac{\pi}{2}, \frac{\pi}{2})$ and $\csc x$ over the interval $(0, \pi)$.

A function $f(x)$ is said to be **concave** over an interval if a chord joining any two points on the graph of the function within the interval lies on or below the graph itself. For most functions, it is sufficient to know that the midpoint of the chord lies on or below the graph. This may be expressed algebraically as $f(\frac{x_1+x_2}{2}) \leq \frac{f(x_1)+f(x_2)}{2}$. Jensen's Inequality then states that for a concave function $f(x)$ over an interval,

$$f\left(\frac{x_1 + x_2 + \cdots + x_n}{n}\right) \leq \frac{f(x_1) + f(x_2) + \cdots + f(x_n)}{n},$$

where $x_1, x_2, \ldots, x_n$ are numbers in this interval.

Example 12.
Prove that $\sin x$ is strictly concave on $[0, \pi]$.

Solution:
Note that $\frac{\sin x_1 + \sin x_2}{2} - \sin \frac{x_1 + x_2}{2} = (\sin \frac{x_1}{2} - \sin \frac{x_2}{2})(\cos \frac{x_1}{2} - \cos \frac{x_2}{2}) \leq 0$
for all x_1 and x_2 in $[0, \pi]$. The last step follows from the fact that $\sin x$ is
increasing while $\cos x$ is decreasing on $[0, \frac{\pi}{2}]$. Equality holds if and only if
$x_1 = x_2$.

Examples of other concave functions are $\sqrt{x}$ over the interval $(0, \infty)$ and
$\cos x$ over the interval $(-\frac{\pi}{2}, \frac{\pi}{2})$. Most results on concave functions may be
obtained from those on convex functions by reversing the inequality signs.

The following (see for instance [8]) is an important result on convex
functions.

Jensen's Inequality.
Let $f(x)$ be a convex function over an interval and let $x_1, x_2, \ldots, x_n$ be
numbers in this interval. Then

$$\frac{f(x_1) + f(x_2) + \cdots + f(x_n)}{n} \geq f\left(\frac{x_1 + x_2 + \cdots + x_n}{n}\right).$$

Equality holds if and only if $x_1 = x_2 = \cdots = x_n$ for strictly convex functions.

Proof:
The case $n = 2$ is just the definition of a convex function. Assuming that
the result holds for some $n \geq 2$, we claim that it also holds for $2n$. Indeed,

$$\frac{f(x_1) + f(x_2) + \cdots + f(x_{2n})}{2n}$$
$$= \frac{\frac{f(x_1)+f(x_2)+\cdots+f(x_n)}{n} + \frac{f(x_{n+1})+f(x_{n+2})+\cdots+f(x_{2n})}{n}}{2}$$
$$\geq \frac{f(\frac{x_1+x_2+\cdots+x_n}{n}) + f(\frac{x_{n+1}+x_{n+2}+\cdots+x_{2n}}{n})}{2}$$
$$\geq f\left(\frac{\frac{x_1+x_2+\cdots+x_n}{n} + \frac{x_{n+1}+x_{n+2}+\cdots+x_{2n}}{n}}{2}\right)$$
$$= f\left(\frac{x_1 + x_2 + \cdots + x_{2n}}{2n}\right).$$

The proof will be complete if we can also show that whenever the result
holds for some $n > 2$, it also holds for $n - 1$. We choose $x_n = \frac{x_1+x_2+\cdots x_{n-1}}{n-1}$.
Then

$$f(x_n) = f\left(\frac{x_1 + x_2 + \cdots + x_{n-1}}{n - 1}\right)$$
$$= f\left(\frac{x_1 + x_2 + \cdots + x_n}{n}\right)$$
$$\leq \frac{f(x_1) + f(x_2) + \cdots + f(x_n)}{n}.$$

Hence $(n-1)f(x_n) \le f(x_1) + f(x_2) + \cdots + f(x_{n-1})$, which is equivalent to

$$\frac{f(x_1) + f(x_2) + \cdots + f(x_{n-1})}{n-1} \ge f\left(\frac{x_1 + x_2 + \cdots + x_{n-1}}{n-1}\right).$$

Using Jensen's Inequality and the fact that a convex function $f(x)$ is continuous, we can deduce that the chord joining any two distinct points on its graph lies entirely on or above the graph. This yields the following result.

Weighted Jensen's Inequality.
Let $f(x)$ be a convex function over an interval and let $x_1, x_2, \ldots, x_n$ be numbers in this interval. Let $a_1, a_2, \ldots, a_n$ be positive numbers. Then

$$\frac{a_1 f(x_1) + a_2 f(x_2) + \cdots + a_n f(x_n)}{a_1 + a_2 + \cdots + a_n} \ge f\left(\frac{a_1 x_1 + a_2 x_2 + \cdots + a_n x_n}{a_1 + a_2 + \cdots + a_n}\right).$$

Equality holds if and only if $x_1 = x_2 = \cdots = x_n$ for strictly convex functions.

Here is a problem from the International Mathematical Olympiad.

Example 13. (2001)
Prove that $\frac{a}{\sqrt{a^2+8bc}} + \frac{b}{\sqrt{b^2+8ca}} + \frac{c}{\sqrt{c^2+8ab}} \ge 1$, where a, b and c are positive real numbers.

First Solution:
The function $\frac{1}{\sqrt{x}}$ is convex over the interval $(0, \infty)$. By the Weighted Jensen's

Inequality, $\dfrac{\frac{a}{\sqrt{a^2+8bc}} + \frac{b}{\sqrt{b^2+8ca}} + \frac{c}{\sqrt{c^2+8ab}}}{a+b+c} \ge \dfrac{1}{\sqrt{\frac{a^3+b^3+c^3+24abc}{a+b+c}}}$. Note that we have

$(a+b+c)^3 = a^3 + b^3 + c^3 + 6abc + 3(b^2c + c^2a + a^2b + bc^2 + ca^2 + ab^2)$, and $b^2c + c^2a + a^2b + bc^2 + ca^2 + ab^2 \ge 6abc$ by the Arithmetic-Mean Geometric-Mean Inequality. Hence $(a+b+c)^3 \ge a^3 + b^3 + c^3 + 24abc$, and the desired inequality follows immediately.

Second Solution:
We have $(a+b+c)^3 \ge a^3 + b^3 + c^3 + 24abc$ As in the First Solution. By Cauchy's Inequality,

$$\begin{aligned}
(a+b+c)^2 &\ge \sqrt{a+b+c}\sqrt{a^3+b^3+c^3+24abc} \\
&\ge \sqrt{a}\sqrt{a^3+8abc} + \sqrt{b}\sqrt{b^3+8abc} + \sqrt{c}\sqrt{c^3+8abc} \\
&= a\sqrt{a^2+8bc} + b\sqrt{b^2+8ca} + c\sqrt{c^2+8ab}.
\end{aligned}$$

Moreover, the product of this expression with $\frac{a}{\sqrt{a^2+8bc}} + \frac{b}{\sqrt{b^2+8ca}} + \frac{c}{\sqrt{c^2+8ab}}$ is greater than or equal to $(a+b+c)^2$. The desired inequality follows immediately.

The concept of convexity can be extended to functions of two or more variables. A function $f(u, v)$ is convex if $\frac{f(u_1,v_1)+f(u_2,v_2)}{2} \ge f(\frac{u_1+u_2}{2}, \frac{v_1+v_2}{2})$.

For instance, $f(u,v) = \frac{u^2}{v}$ is convex for $0 < v < \infty$ since

$$\frac{\frac{u_1^2}{v_1} + \frac{u_2^2}{v_2}}{2} - \frac{(\frac{u_1+u_2}{2})^2}{\frac{v_1+v_2}{2}} = \frac{(v_1+v_2)(u_1^2 v_2 + u_2^2 v_1) - v_1 v_2 (u_1+u_2)^2}{2 v_1 v_2 (v_1 + v_2)}$$

$$= \frac{(u_1 v_2 - u_2 v_1)^2}{2 v_1 v_2 (v_1 + v_2)}$$

$$\geq 0$$

for all positive numbers v_1 and v_2.

Here is another problem from the International Mathematical Olympiad.

Example 14. (1995)
Let a, b and c be positive real numbers such that $abc = 1$. Prove that
$\frac{1}{a^3(b+c)} + \frac{1}{b^3(c+a)} + \frac{1}{c^3(a+b)} \geq \frac{3}{2}$.

First Solution:
Let $x = \frac{1}{a}$, $y = \frac{1}{b}$ and $z = \frac{1}{c}$. Then $xyz = 1$ and

$$\frac{1}{a^3(b+c)} + \frac{1}{b^3(c+a)} + \frac{1}{c^3(a+b)} = \frac{x^2}{y+z} + \frac{y^2}{z+x} + \frac{z^2}{x+y}.$$

Applying Jensen's Inequality to the convex function $f(u,v) = \frac{u^2}{v}$ and then using the Arithmetic-Mean Geometric-Mean Inequality, we have

$$\frac{x^2}{y+z} + \frac{y^2}{z+x} + \frac{z^2}{x+y} \geq 3\left(\frac{(\frac{x+y+z}{3})^2}{\frac{(y+z)+(z+x)+(x+y)}{3}}\right) = \frac{x+y+z}{2} \geq \frac{3}{2}\sqrt[3]{xyz} = \frac{3}{2}.$$

Second Solution:
As in the First Solution, we only need to prove that $\frac{x^2}{y+z} + \frac{y^2}{z+x} + \frac{z^2}{x+y} \geq \frac{3}{2}$. By Cauchy's Inequality, $((y+z)+(z+x)+(x+y))(\frac{x^2}{y+z} + \frac{y^2}{z+x} + \frac{z^2}{x+y}) \geq (x+y+z)^2$. Hence $\frac{x^2}{y+z} + \frac{y^2}{z+x} + \frac{z^2}{x+y} \geq \frac{x+y+z}{2} \geq \frac{3}{2}$.

We are now ready to state our principal result.

The Majorization Inequality.
In $S(n,s)$, let $X > Y$. Let the terms of X and Y be in the domain of a convex function f. Then

$$f(x_1) + f(x_2) + \cdots + f(x_n) \geq f(y_1) + f(y_2) + \cdots + f(y_n).$$

Moreover, equality cannot hold if f is strictly convex.

Proof:
By part (a) of the Fundamental Lemma, we can construct Z in $S(n,s)$ such that $X > Z \geq Y$. We claim that

$$f(x_1) + f(x_2) + \cdots + f(x_n) \geq f(z_1) + f(z_2) + \cdots + f(z_n).$$

Since $z_k = x_k$ for $k \neq i, j$, this is equivalent to $f(x_i) + f(x_j) \geq f(z_i) + f(z_j)$. Let $c = \frac{d}{x_i - x_j}$. Clearly $c > 0$. Since $d \leq x_i - y_i \leq x_i - y_j < x_i - x_j, c < 1$. We have $(1 - c)x_i + cx_j = x_i - d = z_i$ and $cx_i + (1 - c)x_j = xj + d = z_j$. Hence

$$\begin{aligned} f(z_i) + f(z_j) &= f((1 - c)x_i + cx_j) + f(cx_i + (1 - c)x_j) \\ &\leq (1 - c)f(x_i) + cf(x_j) + cf(x_i) + (1 - c)f(x_j) \\ &= f(x_i) + f(x_j). \end{aligned}$$

The desired result now follows from this and part (b) of the Fundamental Lemma. If f is strictly convex, then equality holds if and only if we have $X = Z_0 = Z_1 = \cdots = Z_m = Y$, which is impossible since we have $X > Y$. This completes the proof of the Majorization Inequality.

Jensen's Inequality may be derived as an immediate corollary. We may assume that $x_1 \geq x_2 \geq \cdots \geq x_n$. Let $s = x_1 + x_2 + \cdots + x_n$. In $S(n, s)$, we have $X \geq Y = (\frac{s}{n}, \frac{s}{n}, \ldots, \frac{s}{n})$. If $X > Y$, $f(x_1) + f(x_2) + \cdots + f(x_n) \geq nf(\frac{s}{n})$ or $\frac{f(x_1) + f(x_2) + \cdots + f(x_n)}{n} \geq f(\frac{x_1 + x_2 + \cdots + x_n}{n})$. Equality cannot hold if f is strictly convex. If $X = Y$, then $x_k = \frac{s}{n}$ for $1 \leq k \leq n$, so that $x_1 = x_2 = \cdots = x_n$. Equality clearly holds here.

Example 15.
Let $x_1, x_2, \ldots, x_n$ be positive numbers such that $x_1 + x_2 + \cdots + x_n = 1$. Prove that
$$\frac{x_1}{2 - x_1} + \frac{x_2}{2 - x_2} + \cdots + \frac{x_n}{2 - x_n} \geq \frac{n}{2n - 1}.$$

Solution:
We may assume that $x_1 \geq x_2 \geq \cdots \geq x_n$. Note that for $1 \leq k \leq n, x_k$ lies in $(0, 1]$, on which $\frac{x}{2 - x}$ is strictly convex by Example 11. Note that in $S(n, 1)$, $(x_1, x_2, \ldots, x_n) \geq (\frac{1}{n}, \frac{1}{n}, \ldots, \frac{1}{n})$. By the Majorization Inequality,

$$\frac{x_1}{2 - x_1} + \frac{x_2}{2 - x_2} + \cdots + \frac{x_n}{2 - x_n} \geq n\frac{\frac{1}{n}}{2 - \frac{1}{n}} = \frac{n}{2n - 1}.$$

Equality holds if and only if $x_1 = x_2 = \cdots = x_n$.

Actually, Jensen's Inequality is sufficient here. Such is not the case in the next problem, from the 1992 International Mathematics Tournament of the Towns. See [1] or [3] for the history of this fantastic competition.

Example 16.
Let a and Δ be respectively the length of the side and the area of an equilateral triangle inscribed in a unit circle. A closed polygonal line $A_1 A_2 \ldots A_{51}$ consists of 51 segments, all of length a. It is placed inside the unit circle, and is allowed to intersect itself. Prove that the sum of the areas of the 51 triangles $A_{51} A_1 A_2, A_1 A_2 A_3, \ldots, A_{50} A_{51} A_1$ is not less than 3Δ.

Solution:
Consider a directed segment of length a, initially coinciding with $A_{51}A_1$ and pointing from A_{51} to A_1. For $1 \leq k \leq 51$, rotate it successively about A_k to $A_k A_{k+1}$. In the last step, it is rotated about A_{51} back to $A_{51}A_1$. Since 51 is odd, it is now pointing from A_1 to A_{51} instead. It follows that we must have $\theta_1 + \theta_2 + \cdots + \theta_{51} \geq \pi$, where these angles are $A_{51}A_1A_2, A_1A_2A_3, \ldots,$ $A_{50}A_{51}A_1$ arranged in non-ascending order. Moreover, $0 \leq \theta_k \leq \frac{\pi}{3}$ for $1 \leq k \leq 51$ since the polygonal line lies entirely within the circle. Hence $(\theta_1, \theta_2, \ldots, \theta_{51})$ is majorized by $(\frac{\pi}{3}, \frac{\pi}{3}, \frac{\pi}{3}, \ldots)$. The total area of the 51 triangles is

$$T = \frac{1}{2}a^2(\sin\theta_1 + \sin\theta_2 + \cdots + \sin\theta_{51}).$$

By Example 12, $\sin x$ is strictly concave on $[0, \frac{\pi}{3}]$. The Majorization Inequality yields

$$T \geq \frac{1}{2}a^2\left(\sin\frac{\pi}{3} + \sin\frac{\pi}{3} + \sin\frac{\pi}{3} + \cdots\right) \geq 3\Delta.$$

For an in-depth study of the concept of majorization, see [7].

Exercises

4. Prove that the function $\frac{1}{\sqrt{x}}$ is convex over the interval $(0, \infty)$.

5. Let $x_1, x_2, \ldots, x_n$ be positive numbers such that $x_1 + x_2 + \cdots + x_n = 1$. Use Jensen's Inequality to prove that

$$\frac{x_1}{2 - x_1} + \frac{x_2}{2 - x_2} + \cdots + \frac{x_n}{2 - x_n} \geq \frac{n}{2n - 1}.$$

6. Prove that in $[-\frac{\pi}{6}, \frac{\pi}{6}]$,

$$\cos(2x_1 - x_2) + 2\cos(2x_2 - x_3) + \cdots + \cos(2x_n - x_1)$$

$$\leq \cos x_1 + \cos x_2 + \cdots + \cos x_n.$$

Section 3. Trigonometric Inequalities

This section offers a catalog of inequalities involving sums, products and sums of squares of the angles of a triangle ABC, as well as the corresponding inequalities for the half-angles. Our principal reference is [4].

Problem 1.
Prove that $0 < \sin A + \sin B + \sin C \le \frac{3\sqrt{3}}{2}$.

Solution:
The lower bound is trivial since $\sin x > 0$ on $(0, \pi)$. The sum approaches 0 when one angle approaches $180°$. Note that $\sin x$ is concave on $(0, \pi)$. Hence $\sin A + \sin B + \sin C \le 3 \sin \frac{A+B+C}{3} = \frac{3\sqrt{3}}{2}$ by Jensen's Inequality.

Problem 2.
Prove that $1 < \cos A + \cos B + \cos C \le \frac{3}{2}$.

Solution:
We may assume that $A \le B \le C$. It follows that $B < \frac{\pi}{2}$. Note that we have $\cos A + \cos B + \cos C = 1 + \sin A \sin B - (1 - \cos A)(1 - \cos B)$. If x is an acute angle, $\sin x > \sin^2 x$ and $\cos x > \cos^2 x$ so that $\sin x + \cos x > 1$. It follows that $\sin A \sin B > (1 - \cos A)(1 - \cos B)$ so that $\cos A + \cos B + \cos C > 1$. To prove the upper bound, note that if $C < \frac{\pi}{2}$, then we can use the fact that $\cos x$ is concave on $(0, \frac{\pi}{2})$. We have $\cos A + \cos B + \cos C \le 3 \cos \frac{A+B+C}{3} = \frac{3}{2}$ by Jensen's Inequality. If $C \ge \frac{\pi}{2}$, then $A + B \le \frac{\pi}{2}$. Since $\sin x$ is concave on $(0, \pi)$, $\sin A + \sin B \le 2 \sin \frac{A+B}{2} \le \sqrt{2}$ by Jensen's Inequality. Applying the Arithmetic-Mean Geometric-Mean Inequality, $\sin A \sin B \le \frac{(\sin A + \sin B)^2}{4} \le \frac{1}{2}$. Now $(1 - \cos A)(1 - \cos B) > 0$. It follows that

$$\cos A + \cos B + \cos C < 1 + \sin A \sin B \le \frac{3}{2}.$$

Problem 3.
Prove that $-\infty < \tan A + \tan B + \tan C < \infty$.

Solution:
If one of the angles is slightly less than $\frac{\pi}{2}$, then its tangent is arbitrarily large. If one of the angles is slightly larger than $\frac{\pi}{2}$, then its tangent is arbitrarily small. Thus both bounds are trivial.

Problem 4.
Prove that $\sqrt{3} \le \cot A + \cot B + \cot C < \infty$.

Solution:
If one of the angles is slightly larger than 0, then its cotangent is arbitrarily large. Thus the upper bound is trivial. To prove the lower bound, note that

$$1 + \cos C = 1 - \cos(A + B) = 1 - \cos A \cos B + \sin A \sin B \ge \sin A \sin B.$$

It follows that $\cot A + \cot B = \frac{\sin(A+B)}{\sin A \sin B} \geq \frac{\sin C}{1+\cos C}$. Denote by T the expression $\frac{\sin C}{1+\cos C} + \frac{\cos C}{\sin C}$ so that $\cot A + \cot B + \cot C \geq T$. We also have

$$T = \frac{\sin^2 C + \cos C + \cos^2 C}{\sin C (1 + \cos C)} = \frac{1}{\sin C} + \frac{\sin C}{1 + \cos C}.$$

Hence $2T = 3\frac{\sin C}{1+\cos C} + \frac{1+\cos C}{\sin C} \geq 2\sqrt{3}$ by the Arithmetic-Mean Geometric-Mean Inequality. It follows that $T \geq \sqrt{3}$.

Problem 5.
Prove that $-\infty < \sec A + \sec B + \sec C < \infty$.

Solution:
If one of the angles is slightly less than $\frac{\pi}{2}$, then its secant is arbitrarily large. If one of the angles is slightly larger than $\frac{\pi}{2}$, then its secant is arbitrarily small. Thus both bounds are trivial.

Problem 6.
Prove that $2\sqrt{3} \leq \csc A + \csc B + \csc C < \infty$.

Solution:
If one of the angles is slightly larger than 0, then its cosecant is arbitrarily large. Thus the upper bound is trivial. To prove the lower bound, note that $\csc x$ is convex on $(0, \pi)$. Hence $\csc A + \csc B + \csc C \geq 3 \csc \frac{A+B+C}{3} = 2\sqrt{3}$ by Jensen's Inequality.

Problem 7.
Prove that $0 < \sin A \sin B \sin C \leq \frac{3\sqrt{3}}{8}$.

Solution:
The lower bound is trivial since $\sin x > 0$ on $(0, \pi)$ and the product approaches 0 when one of the angles approaches 0°. It follows from Problem 1 and the Arithmetic-Mean Geometric-Mean Inequality that

$$\sin A \sin B \sin C \leq \frac{\sin A + \sin B + \sin C}{3} \leq \frac{3\sqrt{3}}{8}.$$

Problem 8.
Prove that $-1 < \cos A \cos B \cos C \leq \frac{1}{8}$.

Solution:
The lower bound is trivial since $|\cos x| < 1$ on $(0, \pi)$. For the upper bound, we may assume that triangle ABC is acute. It then follows from Problem 2 and the Arithmetic-Mean Geometric-Mean Inequality that

$$\cos A \cos B \cos C \leq \frac{\cos A + \cos B + \cos C}{3} \leq \frac{1}{8}.$$

Problem 9.
Prove that $-\infty < \tan A \tan B \tan C < \infty$.

Solution:

If one of the angles is slightly less than $\frac{\pi}{2}$, then its tangent is arbitrarily large. If one of the angles is slightly larger than $\frac{\pi}{2}$, then its tangent is arbitrarily small. Thus both bounds are trivial.

Problem 10.

Prove that $-\infty < \cot A \cot B \cot C \leq \frac{\sqrt{3}}{9}$.

Solution:

If one of the angles is slightly larger than 0, then its cotangent is arbitrarily large. If another angle is greater than $\frac{\pi}{2}$, then its cotangent is negative. Thus the lower bound is trivial. To prove the upper bound, we may assume that triangle ABC is acute. Note that $\tan x$ is convex on $(O, \frac{\pi}{2})$. Hence $\tan A + \tan B + \tan C \geq \tan \frac{A+B+C}{3} = 3\sqrt{3}$ by Jensen's Inequality. It is easy to prove that $\tan A + \tan B + \tan C = \tan A \tan B \tan C$. By the identity $\cot x = \frac{1}{\tan x}$, $\cot A \cot B \cot C \leq \frac{\sqrt{3}}{9}$.

Problem 11.

Prove that $-\infty < \sec A \sec B \sec C < \infty$.

Solution:

If one of the angles is slightly less than $\frac{\pi}{2}$, then its secant is arbitrarily large. If one of the angles is slightly larger than $\frac{\pi}{2}$, then its secant is arbitrarily small. Thus both bounds are trivial.

Problem 12.

Prove that $\frac{8\sqrt{3}}{9} \leq \csc A \csc B \csc C < \infty$.

Solution:

The desired result follows from Problem 7 and the identity $\csc x = \frac{1}{\sin x}$.

Problem 13.

Prove that $0 < \sin^2 A + \sin^2 B + \sin^2 C \leq \frac{9}{4}$.

Solution:

The lower bound is trivial. To prove the upper bound, we replace $\sin^2 C$ by $\sin^2(A+B) = \sin^2 A \cos^2 B + \cos^2 A \sin^2 B + 2 \sin A \cos A \sin B \cos B$. Using $\sin^2 A = 1 - \cos^2 A$ and $\sin^2 B = 1 - cos^2 B$, the above expression is equal to $\cos^2 A + \cos^2 B - 2 \cos^2 A \cos^2 B + 2 \sin A \cos A \sin B \cos B$, which is equal to $\cos^2 A + \cos^2 B - 2 \cos A \cos B(\cos A \cos B - \sin A \sin B)$. It follows that $\sin^2 A + \sin^2 B + \sin^2 C = 2 + 2 \cos A \cos B \cos C$. The desired result follows from Problem 8.

Problem 14.

Prove that $\frac{3}{4} \leq \cos^2 A + \cos^2 B + \cos^2 C < 3$.

Solution:
The desired result follows from Problem 13 and $\cos^2 x = 1 - \sin^2 x$.

Problem 15.
Prove that $0 < \tan^2 A + \tan^2 B + \tan^2 C < \infty$.

Solution:
If one of the angles is slightly less than $\frac{\pi}{2}$, then its tangent is arbitrarily large. Thus both bounds are trivial.

Problem 16.
Prove that $1 \le \cot^2 A + \cot^2 B + \cot^2 C < \infty$.

Solution:
If one of the angles is slightly larger than 0, then its cotangent is arbitrarily large. Thus the upper bound is trivial. To prove the lower bound, note that $\cot^2 A + \cot^2 B \ge 2 \cot A \cot B$. Similarly, $\cot^2 B + \cot^2 \ge 2 \cot B \cot C$ and $\cot^2 C + \cot^2 A \ge 2 \cot C \cot A$. All three follow from the Arithmetic-Mean Geometric-Mean Inequality. Adding these three inequalities, we have $\cot^2 A + \cot^2 B + \cot^2 C \ge \cot B \cot C + \cot C \cot A + \cot A \cot B$. It is easy to prove that the last expression is identical to 1.

Problem 17.
Prove that $3 < \sec^2 A + \sec^2 B + \sec^2 C < \infty$.

Solution:
The desired result follows from Problem 15 and $\sec^2 x = \tan^2 x + 1$.

Problem 18.
Prove that $4 \le \csc^2 A + \csc^2 B + \csc^2 C < \infty$.

Solution:
The desired result follows from Problem 16 and $\csc^2 x = \cot^2 x + 1$.

Problem 19.
Prove that $1 < \sin \frac{A}{2} + \sin \frac{B}{2} + \sin \frac{C}{2} \le \frac{3}{2}$.

Solution:
Consider the acute triangle with angles $\frac{\pi - A}{2}$, $\frac{\pi - B}{2}$ and $\frac{\pi - C}{2}$ and use the identity $\sin x = \cos(\frac{\pi}{2} - x)$. The desired result follows from Problem 2.

Problem 20.
Prove that $2 < \cos \frac{A}{2} + \cos \frac{B}{2} + \cos \frac{C}{2} \le \frac{3\sqrt{3}}{2}$.

Solution:
Consider the acute triangle with angles $\frac{\pi-A}{2}$, $\frac{\pi-B}{2}$ and $\frac{\pi-C}{2}$ and use the identity $\cos x = \sin(\frac{\pi}{2} - x)$. The upper bound follows Problem 1. To prove the lower bound, observe that $\sin x$ is concave on $(0, \frac{\pi}{2})$. By the Majorization Inequality, $\sin\frac{\pi-A}{2} + \sin\frac{\pi-B}{2} + \sin\frac{\pi-C}{2} \geq \sin\frac{\pi}{2} + \sin\frac{\pi}{2} + \sin 0 = 2$. The inequality is sharp since each of the three angles is less than a right angle.

Problem 21.
Prove that $\sqrt{3} \leq \tan\frac{A}{2} + \tan\frac{B}{2} + \tan\frac{C}{2} < \infty$.

Solution:
Consider the acute triangle with angles $\frac{\pi-A}{2}$, $\frac{\pi-B}{2}$ and $\frac{\pi-C}{2}$ and use the identity $\tan x = \cot(\frac{\pi}{2} - x)$. The desired result follows Problem 4.

Problem 22.
Prove that $3\sqrt{3} \leq \cot\frac{A}{2} + \cot\frac{B}{2} + \cot\frac{C}{2} < \infty$.

Solution:
If one of the angles is slightly larger than 0, then the cotangent of its half-angle is arbitrarily large. Thus the upper bound is trivial. To prove the lower bound, consider the acute triangle with angles $\frac{\pi-A}{2}$, $\frac{\pi-B}{2}$ and $\frac{\pi-C}{2}$ and use the identity $\cot x = \tan(\frac{\pi}{2} - x)$. Note that $\tan x$ is convex on $(0, \frac{\pi}{2})$. Hence for an acute triangle, $\tan A + \tan B + \tan C \geq 3\tan\frac{A+B+C}{3} = 3\sqrt{3}$ by Jensen's Inequality.

Problem 23.
Prove that $2\sqrt{3} \leq \sec\frac{A}{2} + \sec\frac{B}{2} + \sec\frac{C}{2} < \infty$.

Solution:
Consider the acute triangle with angles $\frac{\pi-A}{2}$, $\frac{\pi-B}{2}$ and $\frac{\pi-C}{2}$ and use the identity $\sec x = \csc(\frac{\pi}{2} - x)$. The desired result follows Problem 6.

Problem 24.
Prove that $6 \leq \csc\frac{A}{2} + \csc\frac{B}{2} + \csc\frac{C}{2} < \infty$.

Solution:
If one of the angles is slightly larger than 0, then the cosecant of its half-angle is arbitrarily large. Thus the upper bound is trivial. To prove the lower bound, consider the acute triangle with angles $\frac{\pi-A}{2}$, $\frac{\pi-B}{2}$ and $\frac{\pi-C}{2}$ and use the identity $\csc x = \sec(\frac{\pi}{2} - x)$. Note that $\sec x$ is convex on $(0, \frac{\pi}{2})$. Hence for an acute triangle, $\sec A + \sec B + \sec C \geq 3\sec\frac{A+B+C}{3} = 6$ by Jensen's Inequality.

Problem 25.
Prove that $0 < \sin\frac{A}{2} \sin\frac{B}{2} \sin\frac{C}{2} \leq \frac{1}{8}$.

Solution:

The lower bound is trivial since $\sin x > 0$ on $(0, \frac{\pi}{2})$. By the Arithmetic-Mean Geometric-Mean Inequality, $\sin \frac{A}{2} \sin \frac{B}{2} \sin \frac{C}{2} \leq \left(\frac{\sin \frac{A}{2} + \sin \frac{B}{2} + \sin \frac{C}{2}}{3} \right)^3$ and the upper bound follows from Problem 19.

Problem 26.

Prove that $0 < \cos \frac{A}{2} \cos \frac{B}{2} \cos \frac{C}{2} \leq \frac{3\sqrt{3}}{8}$.

Solution:

The lower bound is trivial since $\cos x > 0$ on $(0, \frac{\pi}{2})$. By the Arithmetic-Mean Geometric-Mean Inequality, $\cos \frac{A}{2} \cos \frac{B}{2} \cos \frac{C}{2} \leq \left(\frac{\cos \frac{A}{2} + \cos \frac{B}{2} + \cos \frac{C}{2}}{3} \right)^3$ and the upper bound follows from Problem 20.

Problem 27.

Prove that $0 < \tan \frac{A}{2} \tan \frac{B}{2} \tan \frac{C}{2} \leq \frac{\sqrt{3}}{9}$.

Solution:

The lower bound is trivial since $\tan x > 0$ on $(0, \frac{\pi}{2})$. To prove the upper bound, let a, b, c and s denote the side lengths and the semiperimeter of ABC respectively. By the Cosine Law, we have $\cos A = \frac{b^2 + c^2 - a^2}{2bc}$. Hence

$$
\begin{aligned}
\tan^2 \frac{A}{2} &= \frac{\sin^2 \frac{A}{2}}{\cos^2 \frac{A}{2}} \\
&= \frac{1 - \cos A}{1 + \cos A} \\
&= \frac{2bc - (b^2 + c^2 - a^2)}{2bc + b^2 + c^2 - a^2} \\
&= \frac{a^2 - (b - c)^2}{(b + c)^2 - a^2} \\
&= \frac{(a + b - c)(a - b + c)}{(a + b + c)(-a + b + c)} \\
&= \frac{(s - b)(s - c)}{s(s - a)}.
\end{aligned}
$$

It follows from the Arithmetic-Mean Geometric-Mean Inequality that

$$
\begin{aligned}
\tan \frac{A}{2} \tan \frac{B}{2} \tan \frac{C}{2} &= \sqrt{\frac{(s - a)(s - b)(s - c)}{s^3}} \\
&\leq \sqrt{\frac{\left(\frac{(s-a)+(s-b)+(s-c)}{3} \right)^3}{s^3}} \\
&= \frac{\sqrt{3}}{9}.
\end{aligned}
$$

Problem 28.

Prove that $3\sqrt{3} \leq \cot \frac{A}{2} \cot \frac{B}{2} \cot \frac{C}{2} < \infty$.

Solution:
The desired result follows from Problem 27 and the identity $\cot x = \frac{1}{\tan x}$.

Problem 29.
Prove that $\frac{8\sqrt{3}}{9} \le \sec \frac{A}{2} \sec \frac{B}{2} \sec \frac{C}{2} < \infty$.

Solution:
The desired result follows from Problem 26 and the identity $\sec x = \frac{1}{\cos x}$.

Problem 30.
Prove that $8 \le \csc \frac{A}{2} \csc \frac{B}{2} \csc \frac{C}{2} < \infty$.

Solution:
The desired result follows from Problem 25 and the identity $\csc x = \frac{1}{\sin x}$.

Problem 31.
Prove that $\frac{3}{4} \le \sin^2 \frac{A}{2} + \sin^2 \frac{B}{2} + \sin^2 \frac{C}{2} < 1$.

Solution:
Since $\cos A + \cos B + \cos C = 1 - 2\sin^2 \frac{A}{2} + 1 - 2\sin^2 \frac{B}{2} + 1 - \sin^2 \frac{C}{2}$, we
have $\sin^2 \frac{A}{2} + \sin^2 \frac{B}{2} + \sin^2 \frac{C}{2} = \frac{(3-(\cos A + \cos B + \cos C)}{2}$. Problem 2 now yields
the desired result.

Problem 32.
Prove that $2 < \cos^2 \frac{A}{2} + \cos^2 \frac{B}{2} + \cos^2 \frac{C}{2} \le \frac{9}{4}$.

Solution:
The desired result follows from Problem 31 and $\cos^2 x = 1 - \sin^2 x$.

Problem 33.
Prove that $1 \le \tan^2 \frac{A}{2} + \tan^2 \frac{B}{2} + \tan^2 \frac{C}{2} < \infty$.

Solution:
If one of the angles is slightly less than π, then the tangent of its half-angle is
arbitrarily large. Thus the upper bound is trivial. The lower bound follows
from the Rearrangement Inequality in that

$$\tan^2 \frac{A}{2} + \tan^2 \frac{B}{2} + \tan^2 \frac{C}{2} \ge \tan \frac{B}{2} \tan \frac{C}{2} + \tan \frac{C}{2} \tan \frac{A}{2} + \tan \frac{A}{2} \tan \frac{B}{2}.$$

It is easy to prove that the last expression is identical to 1.

Problem 34.
Prove that $9 \le \cot^2 \frac{A}{2} + \cot^2 \frac{B}{2} + \cot^2 \frac{C}{2} < \infty$.

Solution:
If one of the angles is slightly larger than 0, then the cotangent of its half-
angle is arbitrarily large. Thus the upper bound is trivial. The lower bound
follows from the Arithmetic-Mean Geometric-Mean Inequality and Problem
28. We have

$$\cot^2 \frac{A}{2} + \cot^2 \frac{B}{2} + \cot^2 \frac{C}{2} \ge 3\sqrt[3]{\cot^2 \frac{A}{2} \cot^2 \frac{B}{2} \cot^2 \frac{C}{2}} \ge 3\sqrt[3]{(3\sqrt{3})^2} = 9.$$

Problem 35.
Prove that $4 \leq \sec^2 \frac{A}{2} + \sec^2 \frac{B}{2} + \sec^2 \frac{C}{2} < \infty$.

Solution:
The desired result follows from Problem 33 and $\sec^2 x = 1 + \tan^2 x$.

Problem 36.
Prove that $12 \leq \csc^2 \frac{A}{2} + \csc^2 \frac{B}{2} + \csc^2 \frac{C}{2} < \infty$.

Solution:
The desired result follows from Problem 34 and $\csc^2 x = 1 + \cot^2 x$.

Exercises

7. Prove that $\tan A + \tan B + \tan C = \tan A \tan B \tan C$ for any triangle ABC.

8. Let ABC be any triangle.

    (a) Prove that $\cot B \cot C + \cot C \cot A + \cot A \cot B = 1$.
    (b) Prove that $\tan \frac{B}{2} \tan \frac{C}{2} + \tan \frac{C}{2} \tan \frac{A}{2} + \tan \frac{A}{2} \tan \frac{B}{2} = 1$.

9. Let ABC be an acute triangle.

    (a) Prove that $2R + r < s$, where R, r and s are the circumradius, inradius and semiperimeter of ABC.
    (b) Prove that $\sin A + \sin B + \sin C > 2$.

Bibliography

[1] K. Bankov and J. Tabov, We challenge your city! Mathematics Competitions, 2 (1989) 67–73.

[2] G. Hardy, J. Littlewood and G. Polya, "Inequalities", Cambridge University Press, Cambridge, paperback edition, (1988) 260–299.

[3] N. Konstantinov, The Tournament of the Towns, Quantum, 1 (1990) 50–51.

[4] R. Kooistra, *Goniometrische Ongelijkheden in een Driehoek*, Nieuwtydschrift, 45 (1957/58) 108–115.

[5] A. Liu and K. Wu, The Majorization Inequality, Math. Inform. Quart. 4 (1994) 106–111.

[6] A. Liu and K. Wu, The Rearrangement Inequality, Math. Competition 8 (1995) 53–60.

[7] A. W. Marshall and I. Olkin, Inequalities: Theory of Majorization and its Applications, Academic Press, New York, 1979.

[8] I. Niven, Maxima and Minima without Calculus, Mathematical Association of America, Washington, 1981, 93–97.

Solution to Exercises

1. (a) By Corollary 2, $\dfrac{a_2}{a_1} + \dfrac{a_3}{a_2} + \cdots + \dfrac{a_n}{a_{n-1}} + \dfrac{a_1}{a_n} \geq n.$

   (b) By the Arithmetic-Mean Geometric Mean Inequality, we have

   $$\frac{1}{n}\left(\frac{a_2}{a_1} + \frac{a_3}{a_2} + \cdots + \frac{a_n}{a_{n-1}} + \frac{a_1}{a_n}\right) \geq \sqrt[n]{\frac{a_2}{1_1} \cdot \frac{a_3}{a_2} \cdots \frac{a_n}{a_{n-1}} \cdot \frac{a_1}{a_n}} = 1.$$

   This is equivalent to the desired inequality.

2. (a) By Corollary 2, $\dfrac{a_1}{b_1} + \dfrac{a_2}{b_2} + \cdots + \dfrac{a_n}{b_n} \geq n.$

   (b) By the Arithmetic-Mean Geometric Mean Inequality, we have

   $$\frac{1}{n}\left(\frac{a_1}{b_1} + \frac{a_2}{b_2} + \cdots + \frac{a_n}{b_n}\right) \geq \sqrt[n]{\frac{a_1}{b_1} \cdot \frac{a_2}{b_2} \cdots \frac{a_n}{b_n}} = 1.$$

   This is equivalent to the desired inequality.

3. (a) We may assume that $a \geq b \geq c$. Then $\frac{1}{b+c} \geq \frac{1}{c+a} \geq \frac{1}{a+b}$. By the
    Rearrangement Inequality, $\frac{a}{b+c} + \frac{b}{c+a} + \frac{c}{a+b} \geq \frac{b}{b+c} + \frac{c}{c+a} + \frac{a}{a+b}$
    and $\frac{a}{b+c} + \frac{b}{c+a} + \frac{c}{a+b} \geq \frac{c}{b+c} + \frac{a}{c+a} + \frac{b}{a+b}$. Adding these two
    inequalities, we have $2\left(\frac{a}{b+c} + \frac{b}{c+a} + \frac{c}{a+b}\right) \geq \frac{b+c}{b+c} + \frac{c+a}{c+a} + \frac{a+b}{a+b} = 3$,
    which is equivalent to the desired inequality.

    (b) Let $x = b + c$, $y = c + a$ and $z = a + b$. Then $y + z - x = 2a$,
    $z + x - y = 2b$ and $x + y - z = 2c$. Hence

    $$
    \begin{aligned}
    & \frac{a}{b+c} + \frac{b}{c+a} + \frac{c}{a+b} \\
    = {} & \frac{1}{2}\left(\frac{y+z-x}{x} + \frac{z+x-y}{y} + \frac{x+y-z}{z}\right) \\
    = {} & \frac{1}{2}\left(\frac{y}{x} + \frac{x}{y} + \frac{x}{z} + \frac{z}{x} + \frac{z}{y} + \frac{y}{z} - 3\right) \\
    \geq {} & \frac{3}{2}.
    \end{aligned}
    $$

4. Using the Arithmetic-Mean Geometric Mean Inequality twice, we have
    $$\frac{\frac{1}{x_1} + \frac{1}{x_2}}{2} \geq \frac{1}{\sqrt[4]{x_1 x_2}} \geq \frac{1}{\sqrt{\frac{x_1 + x_2}{2}}}$$ for all x_1 and x_2 in $(0, \infty)$.

5. Note that the function $\frac{x}{2-x}$ is convex. Hence

    $$
    \frac{x_1}{2 - x_1} + \frac{x_2}{2 - x_2} + \cdots + \frac{x_n}{2 - x_n} \geq \frac{n\left(\frac{x_1 + x_2 + \cdots + x_n}{n}\right)}{2 - \left(\frac{x_1 + x_2 + \cdots + x_n}{n}\right)} = \frac{n}{2n - 1}.
    $$

6. Rearrange $1, 2, \ldots, n$ as $k_1, k_2, \ldots, k_n$ so that $x_{k_1} \geq x_{k_2} \geq \cdots \geq x_{k_n}$,
    and as $m_1, m_2, \ldots, m_n$ so that

    $$
    2x_{m_1} - x_{m_1 + 1} \geq 2x_{m_2} - x_{m_2 + 1} \geq \cdots \geq 2x_{m_n} - x_{m_n + 1 + 1}.
    $$

    Note that $(2x_{m_1} - x_{m_1 + 1}) + (2x_{m_2} - x_{m_2 + 1}) + \cdots + (2x_{m_n} - x_{m_n + 1 + 1})$
    is equal to $x_{k_1} + x_{k_2} + \cdots + x_{k_n}$. Now $2x_{m_1} - x_{m_1 + 1} \geq 2x_{k_1} - x_{k_1 + 1}$.
    This is by the choice of m_1, $2x_{k_1} - x_{k_1 + 1} \geq x_{k_1}$ by the choice of k_1,
    $(2x_{m_1} - x_{m_1 + 1}) + (2x_{m_2} - x_{m_2 + 1}) \geq (2x_{k_1} - x_{k_1 + 1}) + (2x_{k_2} - x_{k_2 + 1})$
    by the choice of m_2 and $(2x_{k_1} - x_{k_1 + 1}) + (2x_{k_2} - x_{k_2 + 1}) \geq x_{k_1} + x_{k_2}$
    by the choice of k_2, and so on. It follows that $(x_{k_1}, x_{k_2}, \cdots, x_{k_n})$ is
    majorized by $(2x_{m_1} - x_{m_1 + 1}, 2x_{m_2} - x_{m_2 + 1}, \cdots, 2x_{m_n} - x_{m_n + 1 + 1})$. All
    these numbers are in the interval $[-\frac{\pi}{2}, \frac{\pi}{2}]$, over which $\cos x$ is concave.
    The desired inequality follows from the Majorization Inequality.

7. Note that $\tan C = -\tan(A + B) = -\frac{\tan A + \tan B}{1 - \tan A \tan B}$. Hence

    $$
    \begin{aligned}
    \tan A + \tan B + \tan C &= (\tan A + \tan B)\left(1 - \frac{1}{1 - \tan A \tan B}\right) \\
    &= \tan A \tan B \tan C.
    \end{aligned}
    $$

8. (a) We have

$$
\begin{aligned}
&\cot B \cot C + \cot C \cot A + \cot A \cot B \\
&= \frac{\cos B \cos C}{\sin B \sin C} + \frac{\cos C \cos A}{\sin C \sin A} + \frac{\cos A \cos B}{\sin A \sin B} \\
&= \frac{\sin(A+B+C)}{\sin A \sin B \sin C} + 1 \\
&= 1.
\end{aligned}
$$

(b) We have

$$
\begin{aligned}
&\tan\frac{B}{2}\tan\frac{C}{2} + \tan\frac{C}{2}\tan\frac{A}{2} + \tan\frac{A}{2}\tan\frac{B}{2} \\
&= \frac{\sin\frac{B}{2}\sin\frac{C}{2}}{\cos\frac{B}{2}\cos\frac{C}{2}} + \frac{\sin\frac{C}{2}\sin\frac{A}{2}}{\cos\frac{C}{2}\cos\frac{A}{2}} + \frac{\sin\frac{A}{2}\sin\frac{B}{2}}{\cos\frac{A}{2}\cos\frac{B}{2}} \\
&= \frac{-\cos\left(\frac{A}{2}+\frac{B}{2}+\frac{C}{2}\right)}{\cos\frac{A}{2}\cos\frac{B}{2}\cos\frac{C}{2}} + 1 \\
&= 1.
\end{aligned}
$$

9. (a) Let a, b, c and Δ denote the side lengths and the area of ABC respectively. We have $\Delta = \frac{abc}{4R} = \sqrt{s(s-a)(s-b)(s-c)} = rs$. We wish to show that $0 > (2r+r)^2 - s^2 = 4R^2 + 4Rr + r^2 - s^2$. Now

$$
\begin{aligned}
4Rr + r^2 - s^2 &= \frac{abc}{s} + \frac{(s-a)(s-b)(s-c)}{s} - s^2 \\
&= \frac{1}{s}(abc + s^3 - (a+b+c)s^2 \\
&\quad + (bc+ca+ab)s - abc - s^3) \\
&= -\frac{(a+b+c)^2}{2} + bc + ca + ab \\
&= -\frac{a^2+b^2+c^2}{2}.
\end{aligned}
$$

It follows that $32\Delta^2$ times $4R^2 - \frac{a^2+b^2+c^2}{2}$ is equal to

$$
8a^2b^2c^2 - (a^2+b^2+c^2)(-a^4 - b^4 - c^4 + 2b^2c^2 + 2c^2a^2 + 2a^2b^2).
$$

Let $k = b^2+c^2-a^2$, $m = c^2+a^2-b^2$ and $n = a^2+b^2-c^2$. Each is positive since ABC is acute. We have $m+n = a^2$, $n+k = b^2$, $k+m = c^2$, $k+m+n = a^2+b^2+c^2$, $mn = a^4 - b^4 - c^4 + 2b^2c^2$, $nk = b^4 - c^4 - a^4 + 2c^2a^2$ and $km = c^4 - a^4 - b^4 + 2a^2b^2$. It follows that $4R^2 - \frac{a^2+b^2+c^2}{2}$ is equal to $\frac{1}{32\Delta^2}$ of

$$
(m+n)(n+k)(k+m) - (k+m+n)(mn+nk+km).
$$

This simplifies to $-kmn < 0$. Hence $2R + r < s$.

(b) We have $2R(\sin A + \sin B + \sin C) = a+b+c = 2s > 4R+2r > 4R$. Hence $\sin A + \sin B + \sin C > 2$.

Chapter Ten: Polynomial Equations

Section 1. Complex Numbers

A polynomial equation in one variable x is an equation of the form $a_0 X^n + a_1 x^{n-1} + a_2 x^{n-2} + \cdots + a_{n-1} x + a_n = 0$, where $a_0, a_1, a_2, \ldots, a_n$ are real numbers with $a_0 \neq 0$. The degree of this equation is said to be n. An equation of degree 1, or a linear equation, always has a unique real root. For instance, the unique real root of $ax + b = 0$, where $a \neq 0$, is $x = -\frac{b}{a}$.

Consider now an equation $ax^2 + bx + c = 0$ of degree 2, or a quadratic equation, where $a \neq 0$. If $b = c = 0$, then $ax^2 = 0$. Hence $x^2 = 0$, and this equation has $x = 0$ as a double root. If $c = 0$ but $b \neq 0$, then $ax^2 + bx = 0 = x(ax + b)$. Hence either $x = 0$ or $ax + b = 0$. The latter yields $x = -\frac{b}{a}$. So there are still two real roots.

Suppose $b = 0$ but $c \neq 0$. Let us consider a few numerical examples. If $x^2 - 9 = 0 = (x - 3)(x + 3)$, we have two real roots $x = \pm 3$. If we have $x^2 - 3 = 0 = (x - \sqrt{3})(x + \sqrt{3})$, there are two real roots $x = \pm\sqrt{3}$. However, $x^2 + 1 = 0$ has no real roots since $x^2 \geq 0$ for all real numbers x, so that $x^2 + 1 \neq 0$.

The easy way out is to say that $x^2 + 1 = 0$ has no solutions and move on. However, $x^2 + 1 = x^2 - (-1) = (x - \sqrt{-1})(x + \sqrt{-1})$ seems to suggest that the roots are $x = \pm\sqrt{-1}$. The only trouble is that $\sqrt{-1}$ is not a real number.

Nevertheless, it would be nice to have solutions to $x^2 + 1 = 0$. To do so, we must expand our real number system to include numbers such as $\sqrt{x}$ where x is a negative real number. We use the symbol i to denote the imaginary number $\sqrt{-1}$. It is defined by the identity $i^2 = -1$. Thus the roots of $x^2 + 1 = 0$ are $x = \pm i$.

A number of the form $a + bi$ is called a **complex number**. Here a and b are real numbers. We call a the real part and b the imaginary part of the complex number $a + bi$.

Some may cast doubt whether there is any point in introducing numbers which are not real. Even if we are primarily interested in real numbers, there are problems about real numbers which are not easy to solve unless we place the problem in the more general context of complex numbers. At the end, the answers will be complex numbers which are actually real numbers.

The usual rules of arithmetic for real numbers are extended to complex numbers.

Addition/Subtraction Rule.

$$(a + bi) \pm (c + di) = a \pm c + (b \pm d)i.$$

© Springer International Publishing AG 2018
A. Liu, *S.M.A.R.T. Circle Minicourses*, Springer Texts in Education, https://doi.org/10.1007/978-3-319-71743-2_10

Multiplication Rule.

$$(a + bi) \cdot (c + di) = ac - bd + (ad + bc)i.$$

Theorem 1.
If $a + bi = 0$, then $a = 0$ and $b = 0$.

Proof:
Suppose $a + bi = 0$. Then $bi = -a$. Hence $-b^2 = a^2$ so that $a^2 + b^2 = 0$. Now a^2 and b^2 are both non-negative; therefore their sum cannot be zero unless each of them is zero. That is, $a = 0$ and $b = 0$.

Theorem 2.
If $a + bi = c + di$, then $a = c$ and $b = d$.

Proof:
By transposition, $a - c + (b - d)i = 0$. By Theorem 1, $a - c = 0$ and $b - d = 0$; that is, $a = c$ and $b = d$.

Thus in order for two complex numbers be equal it is necessary and sufficient that the real parts be equal and the imaginary parts be equal.

We do not need an explicit rule for division. If we have a fraction with a non-real denominator, we simply convert it into an equivalent fraction with a real denominator. To do this, we introduce a new concept.

Definition.
When two complex numbers differ only in the sign of the imaginary part, they are said to be *conjugate*. Thus the conjugate of $a + bi$ is $a + bi$.

Theorem 3.
Both their sum and the product of two conjugate complex numbers are real.

Proof:
Let the numbers be $a + bi$ and $a - bi$. We have $(a + bi) + (a - bi) = 2a$ and $(a + bi)(a - bi) = a^2 - (-b^2) = a^2 + b^2$. Both are real.

To convert a fraction with a non-real denominator into one with a real denominator, we simply multiply the numerator and the denominator by the conjugate of the denominator. Specifically,

$$
\begin{aligned}
\frac{c + di}{a + bi} &= \frac{(c + di)(a - bi)}{(a + bi)(a - bi)} \\
&= \frac{ac + bd + (ad - bc)i}{a^2 + b^2} \\
&= \left(\frac{ac + bd}{a^2 + b^2}\right) + \left(\frac{ad - bc}{a^2 + b^2}\right)i.
\end{aligned}
$$

Thus we see that the sum, difference, product and quotient of two complex numbers is in each case another complex number.

Example 1.

Evaluate (a) $\dfrac{(2+3i)^2}{2+i}$ and (b) $\dfrac{3+2i}{2-5i} + \dfrac{3-2i}{2+5i}$.

Solution:

(a) The given expression is equal to

$$\frac{4-9+12i}{2+i} = \frac{(-5+12i)(2-i)}{(2+i)(2-i)} = \frac{-10+12+29i}{4+1} = \frac{2}{5} + \frac{29}{5}i.$$

(b) The given expression is equal to

$$\frac{(3+2i)(2+5i) + (3-2i)(2-5i)}{(2+5i)(2-5i)} = \frac{-4+17i-4-17i}{29} = -\frac{8}{29}.$$

Non-real complex numbers such as i are neither positive or negative. Clearly, $i \neq 0$. Suppose $i > 0$. Multiplying both sides by the positive number i, we have $-1 = i^2 > 0$, which is a contradiction. Suppose $i < 0$. Multiplying both sides by the negative number i, we have $-1 = i^2 > 0$ again.

It is useful to notice the successive powers of i: $i^1 = i$, $i^2 = -1$, $i^3 = -i$ and $i^4 = 1$. Since each power is obtained by multiplying the one before it by i, we see that these values go in a cycle of length four.

We now turn our attention to the taking of square roots of complex numbers. For instance, $\sqrt{-3} = \sqrt{3}\sqrt{-1} = \sqrt{3}i$. Thus we do not need a new symbol for $\sqrt{-3}$. It turns out that i is the only one we need.

Since $(-a)(-b) = ab$, we have $\sqrt{(-a)(-b)} = \pm\sqrt{ab}$. It would appear that $\sqrt{-a}\sqrt{-b} = \pm\sqrt{ab}$ also. However, this is not the case. Instead, we have $\sqrt{-a}\sqrt{-b} = \sqrt{a}i\sqrt{b}i = \sqrt{ab}i^2 = -\sqrt{ab}$.

Square Root Rule.

$$\sqrt{a+bi} = \pm\sqrt{\frac{\sqrt{a^2+b^2}+a}{2}} \pm i\sqrt{\frac{\sqrt{a^2+b^2}-a}{2}}.$$

In the two terms, we choose the same sign if $b > 0$ and opposite signs if $b < 0$.

Proof:

Let $\sqrt{a+bi} = x + yi$ where x and y are real numbers. Then we have $a + bi = x^2 - y^2 + 2xyi$. By equating real and imaginary parts, $x^2 - y^2 = a$ and $2xy = b$. Now, $(x^2+y^2)^2 = (x^2-y^2)^2 + (2xy)^2 = a^2+b^2$. It follows that $x^2+y^2 = \sqrt{a^2+b^2}$. From this and $x^2-y^2 = a$, we obtain $x^2 = \frac{\sqrt{a^2+b^2}+a}{2}$ and $y^2 = \frac{\sqrt{a^2+b^2}-a}{2}$. Finally, $x = \pm\sqrt{\dfrac{\sqrt{a^2+b^2}+a}{2}}$ and $y = \pm\sqrt{\dfrac{\sqrt{a^2+b^2}-a}{2}}$.

Example 2.
Find the square roots of $-7 - 24i$.

Solution:
Let $\sqrt{-7 - 24i} = x + yi$, where x and y are real number. Then we have $-7 - 24i = x^2 - y^2 + 2xyi$. By Theorem 2, $x^2 - y^2 = -7$ and $2xy = -24$. Now $(x^2 + y^2)^2 = (x^2 - y^2)^2 + (2xy)^2 = 49 + 576 = 625$. Hence $x^2 + y^2 = 25$. From these equations we have $x^2 = 9$ and $y^2 = 16$, so that $x = \pm 3$ and $y = \pm 4$. Since the product xy is negative we must take $x = 3$ and $y = -4$, or $x = -3$ and $y = 4$. Hence $\sqrt{-7 - 24i} = \pm(-3 + 4i)$.

Expressed as $x + yi$, the complex number is said to be in its standard form. We now introduce its trigonometric form. Let $r = \sqrt{x^2 + y^2}$ and $\theta = \arctan \frac{y}{x}$. Then $x = r \cos \theta$ and $y = \sin \theta$ and the complex number is expressed as $r(\cos \theta + i \sin \theta)$.

Definition.
The positive number $r = \sqrt{x^2 + y^2}$ is called the **modulus** and the angle $\theta = \arctan \frac{y}{x}$ is called the **argument** of the complex number $x + yi$. They are denoted by $\mathrm{mod}(x + yi)$ and $\arg(x + yi)$ respectively.

Theorem 4.
When we multiply two complex numbers, we multiply their moduli and add their arguments.

Proof:
Let the two complex numbers be $r(\cos \theta + i \sin \theta)$ and $s(\cos \phi + i \sin \phi)$. Then their product is $rs(\cos \theta \cos \phi - \sin \theta \sin \phi + i \cos \theta \sin \phi + i \sin \theta \cos \phi)$, which is equal to $rs(\cos(\theta + \phi) + i \sin(\theta + \phi))$.

The following result is an immediate corollary of Theorem 4.

Trigonometric Square Root Rule.
$\sqrt{r(\cos \theta + i \sin \theta)} = \pm \sqrt{r}(\cos \frac{\theta}{2} + i \sin \frac{\theta}{2})$.

Trigonometric Solution to Example 2.
Let $-7 - 24i = r(\cos \theta + i \sin \theta)$. Then $r = \sqrt{(-7)^2 + (-24)^2} = 25$ and $\tan \theta = \frac{-24}{-7}$. Since $\pi < \theta < \frac{3\pi}{2}$, $\cos \theta = -\frac{7}{25}$. It follows that $\sqrt{r} = 5$, $\cos^2 \frac{\theta}{2} = \frac{1 + \cos \theta}{2} = \frac{9}{25}$ and $\sin^2 \frac{\theta}{2} = \frac{1 - \cos \theta}{2} = \frac{16}{25}$. Since $\frac{\pi}{2} < \frac{\theta}{2} < \frac{3\pi}{4}$, $\cos \frac{\theta}{2} = -\frac{3}{5}$ while $\sin \frac{\theta}{2} = \frac{4}{5}$. Hence $\sqrt{-7 - 24i} = \pm(-3 + 4i)$.

Finally, we consider the problem of taking cube roots of complex numbers. Suppose $x = \sqrt[3]{1}$. Then $x^3 = 1$, or $0 = x^3 - 1 = (x - 1)(x^2 + x + 1)$. Therefore, either $x - 1 = 0$, or $x^2 + x + 1 = 0$. The latter yields $x = \frac{-1 \pm \sqrt{3}i}{2}$. Thus 1 has three cube roots. One of them is the real number 1. The other two are conjugate complex numbers. Denote by ω the complex root $-\frac{1}{2} + \frac{\sqrt{3}}{2}i$. Then $\omega^2 = \frac{1}{4} - \frac{\sqrt{3}}{2} - \frac{3}{4} = -\frac{1}{2} - \frac{\sqrt{3}}{2}i$, which is the other complex root.

Observe that the successive positive integral powers of ω are 1, ω and ω^2. These values go in a cycle of length three since $\omega^3 = 1$. We also have $\omega^2 + \omega = -1$ since ω satisfies the equation $x^2 + x + 1 = 0$.

Every real number has three cube roots, two of which are complex. The cube roots of a^3 are those of $a^3 \cdot 1$, and therefore they are a, $a\omega$ and $a\omega^2$. Unless otherwise stated, the symbol $\sqrt[3]{a}$ will always be taken to be the real cube root of the real number a.

A complex number $r(\cos\theta + i\sin\theta)$ also has three cube roots. If one of them is γ, the other two are $\omega\gamma$ and $\omega^2\gamma$. The trigonometric form provides a simple way of computing cube roots.

Trigonometric Cube Root Rule.
The cube roots of $r(\cos\theta + i\sin\theta)$ are

$$\sqrt[3]{r}\left(\cos\frac{\theta}{3} + i\sin\frac{\theta}{3}\right),$$

$$\sqrt[3]{r}\left(\cos\frac{2\pi + \theta}{3} + i\sin\frac{2\pi + \theta}{3}\right)$$

and

$$\sqrt[3]{r}\left(\cos\frac{4\pi + \theta}{3} + i\sin\frac{4\pi + \theta}{3}\right).$$

Example 3.
Find the cube root of $1 + i$.

Solution:
We have $1 + i = \sqrt{2}(\cos 45° + i\sin 45°)$. The three cube roots are

$$\sqrt[6]{2}(\cos 15° + i\sin 15°),$$

$$\sqrt[6]{2}(\cos 135° + i\sin 135°) = -\frac{1}{\sqrt[3]{2}} + \frac{1}{\sqrt[3]{2}}i$$

and

$$\sqrt[6]{2}(\cos 255° + i\sin 255°).$$

Exercises

1. Evaluate $(a - \frac{1+\sqrt{-3}}{2})(a - \frac{1-\sqrt{-3}}{2})$ and $\dfrac{(a+i)^3 - (a-i)^3}{(a+i)^2 - (a-i)^2}$.

2. Find the square root of $-8i$

   (a) without using trigonometry;
   (b) using trigonometry.

3. Find the cube root of $-8i$.

Section 2. Cubic Equations

In Section 1, we have solved the special cases of the quadratic equation $ax^2 + bx + c = 0$ where $bc = 0$. We now solve the general case where $bc \neq 0$. The idea is to reduce this to one of the special cases, via a transformation.

Note that since $a \neq 0$, we can rewrite the equation as $x^2 + \frac{b}{a}x + \frac{c}{a} = 0$. We now complete the square $x^2 + 2\frac{b}{2a} + \frac{b^2}{4a^2} = (x - \frac{b}{2a})^2$. The equation is then transformed into $(x + \frac{b}{2a})^2 - \frac{b^2 - 4ac}{4a^2} = 0$. Regarding $x + \frac{b}{2a}$ as a new variable, the linear term in the new equation has been eliminated. This simplified form of the equation yields solutions immediately, namely, $x = \dfrac{-b \pm \sqrt{b^2 - 4ac}}{2a}$. This is the well-known Quadratic Formula.

The expression $\Delta = b^2 - 4ac$ is called the **discriminant** of the quadratic equation $ax^2 + bx + c = 0$. Its sign determines the natures of the roots. If $\Delta > 0$, we have two distinct real roots $x = \frac{-b \pm \sqrt{\Delta}}{2a}$. If $\Delta = 0$, we have a repeated real root $x = -\frac{b}{2a}$. If $\Delta < 0$, we have conjugate complex roots $x = \frac{-b \pm \sqrt{-\Delta}i}{2a}$.

We now try to eliminate the quadratic term from the general cubic equation $ax^3 + bx^2 + cx + d = 0$. This is equivalent to $x^3 + \frac{b}{a}x^2 + \frac{c}{a}x + \frac{d}{a} = 0$. We complete the cube $x^3 + 3\frac{b}{3a}x^2 + 3\frac{b^2}{9a^2}x + \frac{b^3}{27a^3} = (x + \frac{b}{3a})^3$. The equation is transformed into

$$
\begin{aligned}
0 &= \left(x + \frac{b}{3a}\right)^3 - \frac{b^2 - 3ac}{3a^2}x - \frac{b^3 - 27a^2d}{27a^3} \\
&= \left(x + \frac{b}{3a}\right)^3 - \frac{b^2 - 3ac}{3a^2}\left(x + \frac{b}{3a}\right) - \frac{2b^2 - 9abc + 27a^2d}{27a^3}.
\end{aligned}
$$

Regarding $x + \frac{b}{3a}$ as a new variable, the quadratic term in the new equation has been eliminated. Thus we can regard $x^3 + px + q = 0$ as the general form of the cubic equation.

To solve this equation, we use the substitution $x = y + z$. Then we have $x^3 = y^3 + z^3 + 3yz(y + z)$, and the given equation becomes

$$y^3 + z^3 + (3yz + p)(y + z) + q = 0.$$

At present y and z are any two quantities subject to the condition that their sum is equal to one of the roots of the given equation. If we further suppose that they satisfy the equation $3yz + p = 0$, they are completely determinate. We thus obtain $y^3 + z^3 = -q$ and $yz = -\frac{p}{3}$. It follows that y^3 and z^3 are the roots of the quadratic $t^2 + qt - \frac{p^3}{27} = 0$, which can be solved using the Quadratic Formula.

Putting $y^3 = -\frac{q}{2} + \sqrt{\frac{q^2}{4} + \frac{p^3}{27}}$ and $z^3 = -\frac{q}{2} - \sqrt{\frac{q^2}{4} + \frac{p^3}{27}}$, we obtain the value of x from the relation $x = y + z$. Thus

$$x = \left(-\frac{q}{2} + \sqrt{\frac{q^2}{4} + \frac{p^3}{27}}\right)^{\frac{1}{3}} + \left(-\frac{q}{2} - \sqrt{\frac{q^2}{4} + \frac{p^3}{27}}\right)^{\frac{1}{3}}.$$

The above solution is generally known as Cardan's Solution, as it was first published by him in the Ars Magna, in 1545. Cardan obtained the solution from Tartaglia; but the solution of the cubic seems to have been due originally to Scipio Ferreo, about 1505.

We have seen in Section 1 that each of y^3 and z^3 has three cube roots. Hence it would appear that x has nine values. This, however is not the case. Observe that y and z are to be found from the equations $y^3 + z^3 = -q$ and $yz = -\frac{p}{3}$. However, in the process of solution, the second of these is changed into $y^3 z^3 = -\frac{p^3}{27}$, which also holds if $yz = -\frac{\omega p}{3}$ or $yz = -\frac{\omega^2 p}{3}$. Hence the other six values of x are the roots of the cubic equations $x^3 + \omega p x + q = 0$ and $x^3 + \omega^2 p x + q = 0$.

Hence if y and z denote the values of any pair of cube roots which fulfill this condition, the only other admissible pairs will be ωy and $\omega^2 z$, as well as $\omega^2 y$ and ωz, where ω and ω^2 are the imaginary cube roots of unity. It follows that the roots of the equation are $y + z$, $\omega y + \omega^2 z$ and $\omega^2 y + \omega z$.

Example 4.
Solve the equation $x^3 - 15x - 126 = 0$.

Solution:
Let $x = y + z$. Then $y^3 + z^3 + (3yz - 15)(y + z) - 126 = 0$. Let $3yz - 15 = 0$. Then $y^3 + z^3 = 126$ and $y^3 z^3 = 125$. Hence y^3 and z^3 are the roots of the equation $0 = t^2 - 126t + 125 = (t - 125)(t - 1)$. It follows that $y^3 = 125$ and $z^3 = 1$, so that $y = 5$ and $z = 1$. Thus the roots are $y + z = 5 + 1 = 6$, $\omega y + \omega^2 z = \frac{-5 + 5\sqrt{3}i}{2} + \frac{-1 - \sqrt{3}i}{2} = -3 + 2\sqrt{3}i$ and $\omega^2 y + \omega z = -3 - 2\sqrt{3}i$.

Example 5.
Solve the equation $x^3 - 108x + 432 = 0$.

Solution:
Let $x = y + z$. Then $y^3 + z^3 + (3yz - 108)(y + z) + 432 = 0$. Let $3yz - 108 = 0$. Then $y^3 + z^3 = -432$ and $y^3 z^3 = 46656$. Hence y^3 and z^3 are the roots of the equation $0 = t^2 + 432t + 46656 = (t + 216)^2$, so that $y = z = \sqrt[3]{-216} = -6$. Thus the roots are $y + z = -12$, $y\omega + z\omega^2 = -6(\omega + \omega^2) = 6$ and $y\omega^2 + z\omega = 6$.

The expression $\Delta = \frac{q^2}{4} + \frac{p^3}{27}$ is called the **discriminant** of the cubic equation $x^3 + px + q = 0$. Its sign determines the natures of the roots.

(i) If $\Delta > 0$, then y^3 and z^3 are both real and distinct. Let y and z represent their real cube roots. Then the roots are $y + z$, $\omega y + \omega^2 z$ and $\omega^2 y + \omega z$. The first of these is real, and by substituting for ω and ω^2 the other two become $-\frac{y+z}{2} + \frac{(y-z)\sqrt{3}i}{2}$ and $-\frac{y+z}{2} - \frac{(y-z)\sqrt{3}i}{2}$.

(ii) If $\Delta = 0$, then $y^3 = z^3$. In this case $y = z$, and the roots become $2y$, $y(\omega + \omega^2) = -y$ and $y(\omega^2 + \omega) = -y$.

(iii) If $\Delta < 0$, then y^3 and z^3 are complex numbers. Let r and θ be such that $y^3 = r^3(\cos 3\theta + i \sin 3\theta)$ and $z^3 = r^3(\cos 3\theta - i \sin 3\theta)$. Then we have $y = r(\cos\theta + i \sin\theta)$ and $z = r(\cos\theta - i \sin\theta)$. The roots of the cubic equation are $y + z = 2r\cos\theta$, $\omega y + \omega^2 z = 2r\cos(\theta + \frac{2\pi}{3})$ and $\omega^2 y + \omega z = 2r\cos(\theta + \frac{4\pi}{3})$, which are all real and unequal.

The last case is sometimes called the *Irreducible Case* of Cardan's Solution.

Example 6.
Solve the equation $x^3 - \frac{31}{12}x + \frac{77}{54} = 0$.

Solution:
We have $y^3 + y^3 = -\frac{77}{54}$ while $y^3 z^3 = \frac{31^3}{36^3}$. Hence y^3 and z^3 are the roots of $0 = t^2 + \frac{77}{54}t + \frac{29791}{46656}$. By the Quadratic Formula, we have $y^3 = -\frac{77}{108} + \frac{5\sqrt{3}i}{24}$ and $z^3 = -\frac{77}{108} - \frac{5\sqrt{3}i}{24}$. In trigonometric forms,

$$y^3 = \frac{31\sqrt{31}}{216}(\cos 153.15° + i \sin 153.15°)$$

and

$$z^3 = \frac{31\sqrt{31}}{216}(\cos 153.15° - i \sin 153.15°).$$

Taking cube roots, we may have $y = \frac{\sqrt{31}}{6}(\cos 51.05° + i \sin 51.05°)$ and $z = \frac{\sqrt{31}}{6}(\cos 51.05° + i \sin 51.05°)$. It follows that the roots of the cubic equation are $\frac{\sqrt{31}}{6}(2\cos 51.05°) = \frac{7}{6}$, $\frac{\sqrt{31}}{6}(2\cos(120° + 51.05°)) = -\frac{11}{6}$ and $\frac{\sqrt{31}}{6}(2\cos(240° + 51.05°)) = \frac{2}{3}$.

Exercises

4. Solve the equation $x^3 + 21x + 342 = 0$.

5. Solve the equation $x^3 - \frac{16}{3}x + \frac{128}{27} = 0$.

6. Solve the equation $x^3 - \frac{7}{4}x + \frac{3}{4} = 0$.

Section 3. Quartic Equations

In this section, we give two methods for solving quartic equations. In each case, we transform the general equation $ax^4 + bx^3 + cx^2 + dx + e = 0$ into the simpler form $x^4 + px^2 + qx + r = 0$ by an approach analogous to that used in Section 2. Then we solve an auxiliary cubic equation.

The first method for solving quartic equations was obtained by Ferrari, a pupil of Cardan. Rewrite the equation as $x^4 = -px^2 - qx - r$ and add $x^2 k + \frac{k^2}{4}$ to both sides. We have $(x^2 + \frac{k}{2})^2 = (k-p)x^2 - qx + \frac{k^2}{4} - r$. Now the right side is a square if the discriminant $q^2 - 4(k-p)(\frac{k^2}{4} - r) = 0$. We now use the methods in Section 2 to solve the cubic equation $k^3 - pk^2 - 4rk + (4pr - q^2) = 0$. When one value of k is found, we have $(k-p)x^2 - qx + \frac{k^2}{4} - r = (mx+n)^2$, where $m^2 = k - p$, $2mn = -q$ and $n^2 = \frac{k^2}{4} - r$. From $(x^2 + \frac{k}{2})^2 = (mx+n)^2$, the roots of the quartic equation can be found.

Example 7.
Solve the equation $x^4 + 3x^2 - 2x + 3 = 0$.

Solution:
The auxiliary cubic equation is $k^3 - 3k^2 - 12k + 32 = 0$. By inspection, $k = 4$ is a root. Hence $m^2 = k - 3 = 1$ and $n^2 = \frac{k^2}{4} - 3 = 1$. It follows that $m = \pm 1$ and $n = \pm 1$. Since $mn = 1$ is positive, m and n have the same sign. From $(x^2 + \frac{k}{2})^2 = (mx+n)^2$, we have $x^2 + 2 = x + 1$ or $x^2 + 2 = -x - 1$. The former yields $x = \frac{1}{2} \pm \frac{\sqrt{3}}{2}i$ while the latter yields $x = \frac{1}{2} \pm \frac{\sqrt{11}}{2}i$.

The following method was given by Descartes in 1637. Assume that $x^4 + px^2 + qx + r = 0$ can be factored as $(x^2 + kx + m)(x^2 - kx + n)$. Then by equating coefficients, we have $m + n = p + k^2$, $n - m = \frac{q}{k}$ and $mn = r$. Hence $2n = p + k^2 + \frac{q}{k}$ and $2m = p + k^2 - \frac{q}{k}$. It follows that $4r = (p + k^2)^2 - \frac{q^2}{k^2}$. We now use the methods in Section 2 to solve the cubic equation $(k^2)^3 + 2p(k^2)^2 + (p^2 - 4r)k - q^2 = 0$. When one value of k is found, the values of m and n are determined, and the solution of the quartic equation is obtained by solving the two quadratic equations $x^2 + kx + m = 0$ and $x^- kx + n = 0$.

Example 8.
Solve the equation $x^4 - 2x^2 + 8x - 3 = 0$.

Solution:
The auxiliary cubic equation is $(k^2)^3 - 4(k^2)^2 + 16k^2 - 64 = 0$. By inspection, $k^2 = 4$ is a root. Taking $k = 2$, we have $m + n = 2$ and $n - m = 4$. Hence $m = -1$ and $n = 3$. From $x^2 + 2x - 1 = 0$, we have $-1 \pm \sqrt{2}$. From $x^2 - 2x + 3 = 0$, we have $x = 1 \pm \sqrt{2}i$.

For equations of order higher than four, a search for a method of general solution is a futile task, a deep result due to Galois and Abel. The next result is also stated here without proof.

The Fundamental Theorem of Algebra.
Every polynomial of degree n with complex coefficients has n complex roots.

Theorem 5.
In a polynomial equation with real coefficients, non-real roots occur in conjugate pairs.

Proof:
Suppose that $P(x) = 0$ is a polynomial equation with real coefficients, and suppose that it has a root $a + bi$ with $b \neq 0$. We wish to prove that $a - bi$ is also a root. Now $a + bi + a - bi = 2a$ while $(a + bi)(a - bi) = a^2 + b^2$. Divide $P(x)$ by $x^2 - 2ax + a^2 + b^2$. Let the quotient be $Q(x)$ and the remainder be $rx + s$. Then $Q(x)$ has real coefficients and both r and s are real. Now $P(x) = (x^2 - 2ax + a^2 + b^2)Q(x) + rx + s$. When we put $x = a + bi$, we have $P(x) = 0$ and $x^2 - 2ax + a^2 + b^2 = 0$, so that $rx + s = 0$. By Theorem 2, $ra + s = 0$ and $rb = 0$. Hence $r(a - bi) + s = 0$ also. When we put $x = a - bi$, $x^2 - 2ax + a^2 + b^2 = 0$ and $rx + s = 0$, so that $P(x) = 0$. It follows that $a - bi$ is indeed a root of $P(x) = 0$.

Theorem 6.
Every polynomial with real coefficients can be decomposed into factors with real coefficients which are either linear or quadratic.

Proof:
Let $P(x)$ be a polynomial with real coefficients of degree n. By the Fundamental Theorem of Algebra, the equation $P(x) = 0$ has n complex roots. Each real root corresponds to a linear factor of $P(x)$ with real coefficients. The non-real roots come in conjugate pairs by Theorem 5, and each pair corresponds to a quadratic factor of $P(x)$ with real coefficients. Since all the roots are accounted for, there are no other factors.

Theorem 6 is a result involving only real numbers, in both its hypothesis and its conclusion, but is very hard to prove without placing it in the more general context of complex numbers.

Exercises

7. Solve the equation $x^4 - 10x^2 - 20x - 16 = 0$ By Ferrari's method.

8. Solve the equation $x^4 - 3x^2 - 6x - 2 = 0$ by Descartes' method.

9. Let a and b be rational numbers such that $\sqrt{b}$ is irrational. Let $P(x)$ be a polynomial with rational coefficients. Prove that if $P(a + \sqrt{b}) = 0$, then $P(a - \sqrt{b}) = 0$.

Bibliography

[1] L. E. Dickson, Elementary Theory of Equations, John Wiley & Sons, New York, 1914.

[2] H. S. Hall and R. S. Knight, Higher Algebra, MacMillan, London, 1891.

[3] H. W. Turnbull, Theory of Equations, 4th ed., Oliver & Boyd, Edinburgh, 1947.

[4] J. V. Uspensky, Theory of Equations, McGraw-Hill, New York, 1948.

Solution to Exercises

1. (a) We have $a^2 - (\frac{1+\sqrt{3}i}{2} + \frac{1-\sqrt{3}i}{2})a + \frac{1^2 - (\sqrt{3}i)^2}{2^2} = a^2 - a + 1$.

    (b) We have $\frac{a^3 + 3a^2 i - 3a - i - a^3 + 3a^2 i + 3a - i}{a^2 + 2ai - 1 - a^2 + 2ai + 1} = \frac{3a^2 - 1}{2a}$.

2. (a) Let $\sqrt{-8i} = x + yi$. Then $-8i = x^2 - y^2 + 2xyi$. Hence $x^2 - y^2 = 0$ and $xy = -4$. From $x^2 - y^2 = 0$, either $y = x$ or $y = -x$. Since $xy = -4$, we must have $y = -x$. Hence $-x^2 = -4$ so that $x = \pm 2$ and $y = \mp 2$. It follows that $\sqrt{-8i} = \pm 2(1 - i)$.

    (b) In the trigonometric form, $-8i = 8(\cos \frac{3\pi}{2} + i \sin \frac{3\pi}{2})$. Hence $\sqrt{-8i} = \pm 2\sqrt{2}(\cos \frac{3\pi}{4} + i \sin \frac{3\pi}{4}) = \pm 2(1 - i)$.

3. One of the cube roots of $-8i$ is $2(\cos \frac{\pi}{2} + i \sin \frac{\pi}{2} = 2i$. The others are $2i\omega = -\sqrt{3} - i$ and $2i\omega^2 = \sqrt{3} - i$.

4. We have $y^3 + z^3 = -342$ while $y^3 z^3 = -\frac{21^3}{27} = -343$. From the quadratic equation $0 = t^2 + 342t - 343 = (t - 1)(t + 343)$, we have $y = \sqrt[3]{1} = 1$ and $z = \sqrt[3]{-343} = -7$. Hence the roots of the cubic equation are $y + z = -6$, $\omega y + \omega^2 z = -\frac{1}{2} + \frac{\sqrt{3}}{2}i + \frac{7}{2} + \frac{7\sqrt{3}}{2}i = 3 + 4\sqrt{3}i$ and $\omega^2 y + \omega z = 3 - 4\sqrt{3}i$.

5. We have $y^3 + z^3 = -\frac{128}{27}$ while $y^3 z^3 = -\frac{1}{27}(-\frac{16}{3}) = \frac{4096}{729}$. From the quadratic equation $0 = t^2 + \frac{128}{27}t + \frac{4096}{729} = (t + \frac{64}{27})^2$, we have $y = -\frac{4}{3}$ and $z = -\frac{4}{3}$. Hence the roots of the cubic equation are $y + z = 2y = -\frac{8}{3}$, $\omega y + \omega^2 z = -y = \frac{4}{3}$ and $\omega^2 y + \omega z = -y = \frac{4}{3}$.

6. We have $y^3 + y^3 = -\frac{3}{4}$ while $y^3 z^3 = \frac{7^3}{12^3}$. Hence y^3 and z^3 are the roots of $0 = t^2 + \frac{3}{4}t + \frac{343}{1728}$. By the Quadratic Formula, we have $y^3 = -\frac{3}{8} + \frac{5\sqrt{3}i}{36}$ and $z^3 = -\frac{3}{8} - \frac{5\sqrt{3}i}{36}$.

In trigonometric forms,

$$y^3 = \frac{\sqrt{1029}}{72}(\cos 147.3° + i \sin 147.3°)$$

and

$$z^3 = \frac{\sqrt{1029}}{72}(\cos 147.3° - i \sin 147.3°).$$

Taking cube roots, we may have $y = \sqrt[6]{\frac{1029}{81}}(\cos 49.1° + i \sin 49.1°)$ and $z = \sqrt[6]{\frac{1029}{81}}(\cos 49.1° + i \sin 49.1°)$. It follows that the roots of the cubic equation are $\sqrt[6]{\frac{1029}{81}}(2 \cos 49.1°) = 1$, $\sqrt[6]{\frac{1029}{81}}(2 \cos(120° + 49.1°)) = -\frac{3}{2}$ and $\sqrt[6]{\frac{1029}{81}}(2 \cos(240° + 49.1°)) = \frac{1}{2}$.

7. The auxiliary cubic equation is $k^3 + 10k^2 + 64k + 240 = 0$. By inspection, $k = -6$ is a root. Hence $m^2 = k - (-10) = 4$ and $n^2 = \frac{k^2}{4} - (-16) = 25$. It follows that $m = \pm 2$ and $n = \pm 5$. Since $mn = -(-10)$ is positive, m and n have the same sign. From $(x^2 + \frac{k}{2})^2 = (mx + n)^2$, we have $x^2 - 3 = 2x + 5$ or $x^2 - 3 = -2x + 5$. The former yields $x = 4$ or $x = -2$ while the latter yields $x = -1 \pm i$.

8. The auxiliary cubic equation is $(k^2)^3 - 6(k^2)^2 + 17k^2 - 36 = 0$. By inspection, $k^2 = 4$ is a root. Taking $k = 2$, we have $m + n = 1$ and $n - m = -3$. Hence $m = 2$ and $n = -1$. From $x^2 + 2x + 2 = 0$, we have $-1 \pm i$. From $x^2 - 2x - 1 = 0$, we have $x = 1 \pm \sqrt{2}$.

9. We first prove that if $m + \sqrt{n} = 0$, then $m = n = 0$. This is because a rational number cannot be equal to an irrational number. Now $a + \sqrt{b} + a - \sqrt{b} = 2a$ while $(a + \sqrt{b})(a - \sqrt{b}) = a^2 - b$. Divide $P(x)$ by $x^2 - 2ax + a^2 - b$. Let the quotient be $Q(x)$ and the remainder be $rx + s$. Then $Q(x)$ has rational coefficients and both r and s are rational. Now $P(x) = (x^2 - 2ax + a^2 + b^2)Q(x) + rx + s$. When we put $x = a + \sqrt{b}$, we have $P(x) = 0$ and $x^2 - 2ax + a^2 - b = 0$, so that $rx + s = 0$. If $r = 0$, then $s = 0$ also. We may as well assume that $r \neq 0$. Since $\sqrt{b}$ is irrational, $b \neq 0$, so that $r\sqrt{b}$ is irrational. Thus $ra + s = 0$ and $r\sqrt{b} = 0$. Hence $r(a - \sqrt{b}) + s = 0$ also. When we put $x = a - \sqrt{b}$, $x^2 - 2ax + a^2 - b = 0$ and $rx + s = 0$, so that $P(x) = 0$. It follows that $a - \sqrt{b}$ is indeed a root of $P(x) = 0$.